4/60

European and North
American Bats
TS-289

MOST IMPORTANT MEASUREMENTS

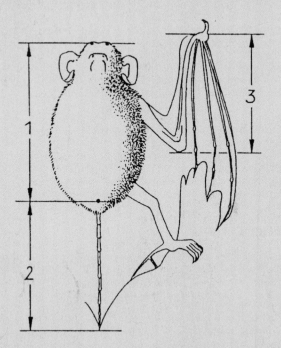

1: Head-body Length
2: Tail Length
3: Forearm Length

4: Ear Length (ear)
5a: Tragus Length
5b: Tragus Width

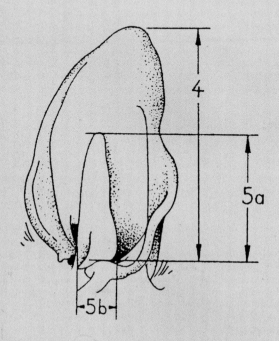

6a: Thumb Length
6b: Length of the Thumb Claw

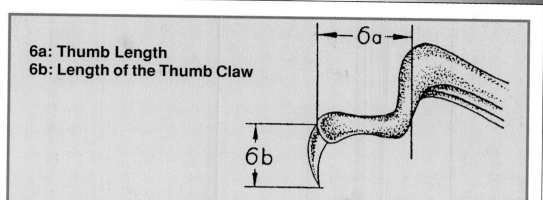

7: Wingspan
8: Length of the Fifth Finger (more correctly: length of the fifth metacarpal and the fifth finger)

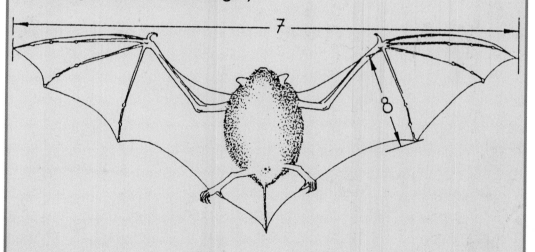

9: Condylobasal Length

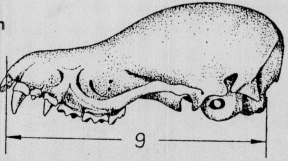

Long-eared Bat
capturing prey.

THE BATS OF EUROPE & NORTH AMERICA

KNOWING THEM—IDENTIFYING THEM—PROTECTING THEM

WILFRIED SCHOBER & ECKARD GRIMMBERGER
TRANSLATOR: WILLIAM CHARLTON

CUSTOMARY U.S. MEASURES AND EQUIVALENTS

LENGTH

1 INCH (IN)		= 2.54 CM
1 FOOT (FT)	= 12 IN	= .3048 M
1 YARD (YD)	= 3 FT	= .9144 M
1 MILE (MI)	= 1760 YD	= 1.6093 KM
1 NAUTICAL MILE	= 1.152 MI	= 1.853 KM

AREA

1 SQUARE INCH (IN^2)		= 6.4516 CM^2
1 SQUARE FOOT (FT^2)	= 144 IN^2	= .093 M^2
1 SQUARE YARD (YD^2)	= 9 FT^2	= .8361 M^2
1 ACRE	= 4840 YD^2	= 4046.86 M^2
1 SQUARE MILE(MI^2)	= 640 ACRE	= 2.59 KM^2

WEIGHT

1 OUNCE (OZ)	= 437.5 GRAINS	= 28.35 G
1 POUND (LB)	= 16 OZ	= .4536 KG
1 SHORT TON	= 2000 LB	= .9072 T
1 LONG TON	= 2240 LB	= 1.0161 T

VOLUME

1 CUBIC INCH (IN^3)		= 16.387 CM^3
1 CUBIC FOOT (FT^3)	= 1728 IN^3	= .028 M^3
1 CUBIC YARD (YD^3)	= 27 FT^3	= .7646 M^3
1 FLUID OUNCE (FL OZ)		= 2.957 CL
1 LIQUID PINT (PT)	= 16 FL OZ	= .4732 L
1 LIQUID QUART (QT)	= 2 PT	= .946 L
1 GALLON (GAL)	= 4 QT	= 3.7853 L
1 DRY PINT		= .5506 L
1 BUSHEL (BU)	= 64 DRY PT	= 35.2381 L

METRIC MEASURES AND EQUIVALENTS

LENGTH

1 MILLIMETER (MM)		= .0394 IN
1 CENTIMETER (CM)	= 10 MM	= .3937 IN
1 METER (M)	= 1000 MM	= 1.0936 YD
1 KILOMETER (KM)	= 1000 M	= .6214 MI

AREA

1 SQ CENTIMETER (CM^2)	= 100 MM^2	= .155 IN^2
1 SQ METER (M^2)	= 10,000 CM^2	= 1.196 YD^2
1 HECTARE (HA)	= 10,000 M^2	= 2.4711 ACRES
1 SQ KILOMETER (KM^2)	= 100 HA	= .3861 MI^2

WEIGHT

1 MILLIGRAM (MG)		= .0154 GRAIN
1 GRAM (G)	= 1000 MG	= .0353 OZ
1 KILOGRAM (KG)	= 1000 G	= 2.2046 LB
1 TONNE (T)	= 1000 KG	= 1.1023 SHORT TONS
1 TONNE		= .9842 LONG TON

VOLUME

1 CUBIC CENTIMETER (CM^3)	= .061 IN^3	
1 CUBIC DECIMETER (DM^3)	= 1000 CM^3	= .353 FT^3
1 CUBIC METER (M^3)	= 1000 DM^3	= 1.3079 YD^3
1 LITER (L)	= 1 DM^3	= .2642 GAL
1 HECTOLITER (HL)	= 100 L	= 2.8378 BU

TEMPERATURE

CELSIUS°	= 5/9 (F° − 32°)
FAHRENHEIT°	= 9/5 (C° + 32°)

This book was originally published by Franckh'sche Verlagshandlung. They own the original copyright in the German language. This English edition expanded and enlarged the original German edition. The new material starts on page 210 to emphasize the North American bats. Twelve additional color photographs have been supplied by the Bat Conservation International, Inc. All appear between pages 212 and 224. All were taken by Dr. Merlin D. Tuttle.

Copyright is claimed by T.F.H. Publications for the English translation and the additional material.

© 1997 by T.F.H. Publications, Inc.

Distributed in the UNITED STATES to the Pet Trade by T.F.H. Publications, Inc., One T.F.H. Plaza, Neptune City, NJ 07753; distributed in the UNITED STATES to the Bookstore and Library Trade by National Book Network, Inc. 4720 Boston Way, Lanham MD 20706; in CANADA to the Pet Trade by H & L Pet Supplies Inc., 27 Kingston Crescent, Kitchener, Ontario N2B 2T6; Rolf C. Hagen Inc., 3225 Sartelon St. Laurent-Montreal Quebec H4R 1E8; in CANADA to the Book Trade by Vanwell Publishing Ltd., 1 Northrup Crescent, St. Catharines, Ontario L2M 6P5 ; in ENGLAND by T.F.H. Publications, PO Box 15, Waterlooville PO7 6BQ; in AUSTRALIA AND THE SOUTH PACIFIC by T.F.H. (Australia), Pty. Ltd., Box 149, Brookvale 2100 N.S.W., Australia; in NEW ZEALAND by Brooklands Aquarium Ltd. 5 McGiven Drive, New Plymouth, RD1 New Zealand; in Japan by T.F.H. Publications, Japan—Jiro Tsuda, 10-12-3 Ohjidai, Sakura, Chiba 285, Japan; in SOUTH AFRICA by Lopis (Pty) Ltd., P.O. Box 39127, Booysens, 2016, Johannesburg, South Africa. Published by T.F.H. Publications, Inc.

MANUFACTURED IN THE
UNITED STATES OF AMERICA
BY T.F.H. PUBLICATIONS, INC.

FOREWORD

About 900 species of bats and flying foxes inhabit the world. Their main areas of occurrence are in the warm regions. That's why only 30 species occur in Europe. Even today, few people have any detailed knowledge about these peculiar creatures, which flit silently through the night, "see" with their ears, fly with their "hands," and sleep hanging by the toes of their hind legs. Throughout the history of civilization we find the most absurd reports about these mammals, and even today the most diverse media use the term "vampire bat" to induce fear in people.

There is no doubt that the many peculiarities in the life style of bats contribute to their being regarded with so many prejudices, being despised, or exterminated in the past and even today. Yet, it is not superstition and fear on the part of people that threaten the existence of these animals today; rather it is the advance of civilization and industrialization. It is therefore imperative that we strongly support the preservation of these beneficial animals.

The pursuit of bats requires a knowledge of the individual species. Although the number of species that occur in Europe seems small, they are not always easy to distinguish by the lay person, and in some species not even by the expert. Because there is little readily comprehensible literature available to those with an interest in bats, we want here to take a look at the fascinating results of modern bat research and to present the European species in text and illustrations to a broad readership. Finally, the diagnostic key presented in this book should help to identify the species that are found.

It is of special interest to us to give our warm thanks to all our friends and colleagues for their expert advice, for providing photographs, literature, and collected material, as well as for their support on the many joint excursions. Special thanks go to Asst. Prof. Dr. J. Gaisler (Brno), Asst. Prof. Dr. H. Hackethal (Berlin), Prof. Dr. v. Helversen, Dr. K. Heller, and R. Weid (Erlangen), as well as RNDr. Zd. Bauerová (Brno), Graduate Secondary School Teacher A. Benk (Hannover), M. Bogdanowicz (Bialowieza), Ing. J. C#erven+ (Prague), B. Evtimov (Pe+utera), J. Gebhard (Basel), Dr. J. Haensel (Berlin), G. Heise (Prenzlau), Dr. U. Jüdes (Kiel), M. Masing (Tartu), W. Oldenburg (Waren), Zb. Urbanczyk (Poznán), Prof. Dr. Y. Tupinier (Caluire), and M. Wilhelm (Dresden).

We also wish to give our warmest thanks to Mrs. T. Schneehagen (Leipzig), who prepared the illustrations with such great skill.

Wilfried Schober
Eckhard Grimmberger

CONTENTS

Common Pipistrelle in abandoned hole of the Greater Spotted Woodpecker.

ON THE LIFE OF BATS

BATS—VAMPIRES, DEVILS, OR DEITIES?

Even today, in our modern and enlightened world, bats continue to arouse fear, terror, or repugnance. When we search for the causes of this, it turns out that few people have any detailed knowledge about the appearance and the habits of these animals. As is true of many nocturnal animals, such as owls, bats have stimulated the human imagination since time immemorial. For centuries a negative role was ascribed to them. In ancient Rome, for example, Divus Basilius wrote: "The nature of the bat is that of a blood relative of the devil." In the Baroque Age, as well, the bat was regarded as a symbol of the Antichrist. Furthermore, Christian art often depicts the Devil and his hellish followers with bats' wings, whereas angels have birds' wings.

The Spanish painter Goya used bats as a symbol of the ominous and the stupid. Even today, Count Dracula continues to wander through films and television in the form of a blood-sucking vampire, often taking the form of a bat.

Bats have had (and continue to have) supernatural powers ascribed to them. They were components of amulets among the medicine men of various primitive peoples. In the books of the ancient Arab physicians, we find numerous recipes in which whole bats or parts of them are used, and they frequently were components of the "medicines" of the European apothecaries of the Middle Ages as well. In India, even today live flying foxes are offered in the bazaars for medicinal purposes. Their skin is pulled off and placed raw on diseased body parts.

Because up to about 50 years ago no one knew anything about how bats navigate with the aid of ultrasound, it was assumed that bats could see in the dark. From this were also derived medicinal recommendations, such as can be read in Albertus Magnus's (13th century) book *The Wonders of the World*: "If you wish to see something in the deepest night and nothing is to be more hidden from you than as during the day, then anoint your face with the blood of a bat and everything will happen to you as I have said."

The ability of bats to fly has caused them to be regarded for millennia not as mammals, but as birds. In the Bible (Leviticus, Chapter 11:13) the faithful are admonished as follows: "... And these you should regard as an abomination among the birds; they shall not be eaten, they are an abomination: the eagle, the vulture, the buzzard ... the hoopoe, and the bat." Bats are thus classified with the "unclean" birds, whose consumption is forbidden.

Enlightened people, on the other hand, such as the brilliant painter and scientist Leonardo da Vinci, saw bats through different eyes. In the construction plans for his flying machine he indicated that the bat wing served as his model. One of his construction sketches clearly showed that the shape of the wings of his flying machine, as well as their braces acting as fingers, resembled the bat wing.

Cities, such as Valencia in Spain, found bats worthy of being placed on their coat of arms. Elector Friedrich the Wise of Kursachsen awarded the painter Lucas Cranach the Elder a letter of heraldry and thus the right to bear: "... a plain shield, upon it a black snake, with two black bat wings in the center ..." as a coat of arms.

deities of the Mayas was represented as a person with spread bat wings and a bat's head. Depictions of the same are found on stone columns and clay pots that were excavated near 2000-year-old temples. In the hieroglyphics of the Mayas, the symbol for the bat occurs frequently. In China and Japan bats are a symbol of luck. The Chinese word "fu" means both bat and luck. A widespread talisman

In Christian art devils are portrayed with bats wings (Rila Monastery, Bulgaria).

Outside Europe, bats often had a completely different and considerably more positive meaning for people. In the ancient cultures of Central America, bats played an important role in the history of religion. One of the

is a kind of coin showing a tree with roots and branches as the symbol of life. Five bats ("wu fu") encircle it with their wings. The bearer of the talisman is promised a long life, riches, health, luck, and an easy death.

Today bats are also enjoying a slowly increasing benevolent interest. With many people,

Medallion on a Chinese robe from the eighteenth century. It shows an ornamental design, in which five bats encircle a tree of life. They embody the concepts of long life, wealth, health, luck, and an easy death.

however, repulsion or indifference still prevails. The unfounded fear that bats could fly into a woman's hair cannot be stamped out. As before, bats are killed, poisoned, or smoked out. This shows how little is known even today about how harmless bats are and how they are even very beneficial as "biological pest control experts." They have at their disposal many abilities and traits that are unique in the animal kingdom; they are the only mammals that can fly actively and use their "hands" to do so. They "see" with their ears, hang upside down by the claws of their hind feet to sleep, and can, in comparison to other small mammals, attain the truly "ancient" age of 30 years.

Bats have inhabited the earth for about 50 million years. In this time they have conquered many ecological niches and have been able to evolve without disturbance. But in the last 40 years humans, particularly in the highly industrialized countries, as a result of accelerating changes to the environment, have threatened the existence of many plant and animal species, and particularly that of our bats. To ensure that interested and enlightened groups of people will intervene widely for the preservation of these lovable and beneficial animals, we are all going to have to do a better job of communicating to the general public how valuable they actually are.

A RESIDENT OF EARTH FOR 50 MILLION YEARS

In the book *Historia Animalium*, which was written in the 16th century by the naturalist Konrad Gesner in Zurich, we read: "The bat is an animal intermediate between the bird and the mouse, and thus can justly be called a flying mouse, although it can be numbered neither among the birds nor the mice, because it combines both of these forms."

It took a long time before bats were allocated their scientifically grounded place in the animal kingdom. Upon superficial examination, we find various characters that apply equally to bats and mice. With many European species, in particular, the body sizes, coat color, or ear shapes are very similar to those of mice. The comparison with mice, which crops up repeatedly, is also reflected in their names; for example, Greater Mouse-eared Bat (*Myotis myotis*) or in the French "*chauve-souris*" (naked mouse.)

Bats, however, are not flying mice at all, but rather an independent order within the mammals that have been given the scientific name Chiroptera (= hand fliers). The most important common character of this mammalian order, the most species rich after the rodents, is the transformation of the front extremities into an organ of flight.

About 250 years ago, the Swede Carolus Linnaeus was the first to begin with, at that time, a comprehensive classification of animals, plants, and rocks. Starting from the concept of "species" in his work *Systema Naturae* (1735), he classified all living creatures according to their presumed family relationships. Linnaeus, who knew of only six bat species at the time, two of them from Europe, initially classified them with the carnivores, because he was guided by the characteristics of the teeth. Because new criteria constantly came to light, it is not surprising to us that 30 years later Linnaeus classified the bats as relatives of the apes, proceeding from the assumption that the location of the mammary glands on the chest was an important character. Extensive collecting trips, which were carried out in the following 100 years, contributed enormously to the discovery of new species. Thus, in 1865, the zoologist Koch mentions more than 300 species in his work "*The essentials of the Chiropterids with the separate description of the bats occurring in the countryside adjoining the Duchy of Nassau.*"

The discovery and description of new species has yet to come to an end. The discovery, about 15 years ago, of a new bat species caused a great sensation, because it is the only representative of a separate new bat family. This bat, with the

Vampire Bat (*Desmodus rotundus*).

complicated scientific name *Craseonycteris thonglongyai*, with a weight of approximately two grams and a head-trunk length of 29 to 33 millimeters, is, besides the Etruscan Shrew (*Suncus etruscus*), the smallest mammal in the world. Some insects, such as our native Stag Beetle (*Lucanus cervus*), the males of which grow up to 80 millimeters long, are clearly larger and heavier. The largest representative of the order of the bats is the fruit-eating Kalong (*Pteropus vampyrus*), which belongs to the same suborder as the flying foxes. With a wingspan of up to 1.7 meters and a weight of 900 grams, it is quite an imposing animal. Among the European species, the Common Pipistrelle (weighing 3 to 8 grams, with a wingspan of about 20 centimeters) and the Greater Mouse-eared Bat (weighing 20 to 40 grams, with a wingspan up to 41 centimeters) represent the extremes.

Only the rare Greater Noctule (weighing 41 to 76 grams, with a wingspan of 41 to 46 centimeters) exceeds the Greater Mouse-eared Bat in size. The evolution of this mammalian order, which extended over millions of years, led to an enormous diversity of forms. Scientists are not in agreement about the number of bat species that currently exist in the world. The total ranges from 800 to 950 species. The main reason for this

Facing page: Flying Fox (*Rousettus stresemanni*). The large eyes argue in favor of good seeing ability in twilight by the Megachiroptera.

wealth of species can be found primarily in the acquisition of the capability of flight. The same reasoning also explains the diversity of birds and insects. As a result of their almost exclusively nocturnal habits and the mastery of the air, it was possible for bats to exploit living spaces and food sources (ecological niches) that were not available to the other mainly terrestrial mammals or the predominantly diurnal birds. All of the bats living in Europe feed on insects. As a rule, they do not set out on their hunting flights until dusk or after dark. Among the native insectivorous birds, only the Nightjar (*Caprimulgus*), also known as the Goatsucker, hunts its insect prey at dusk.

The dietary spectrum of tropical bats is considerably more varied. There are fruit-eating species and "flower bats," which feed on pollen, thus contributing to the pollination of the flowers. Certain species have specialized in hunting smaller bats, mice, and other animals, and there are species that feed on fishes or frogs.

Widely known, at least by name, are the vampire bats that live in Central and South America, the most familiar representative of which is *Desmodus rotundus*. These bats make a small skin wound with their sharp teeth on domestic animals, as well as humans, and then lick up the welling blood. The loss of blood in itself is slight and, as a rule, harmless. Nevertheless, vampires can be dangerous because of the transmission of rabies and other diseases.

The majority of the aforementioned dietary specialists are also nocturnal, thus avoiding competitors and enemies. Certain bats, however, such as the African Greater Leaf-nosed Bat (*Lavia frons*), also hunt insects during the day.

Taxonomists divide the Chiroptera into 18 families. Some families contain only one species, whereas about 320 species belong to the common bat family, Vespertilionidae. In general, bats can be divided into two large suborders. The first is the Megachiroptera, primarily large forms that are known by the common names "flying foxes" or "megabats." Approximately 175 species have been described so far. They inhabit the tropics and subtropics of the Old World. The second suborder is the Microchiroptera, predominantly small forms that are generally known as "bats" or "microbats."

The ancestors of the bats were four-legged mammals—the possession of an organ of flight is a secondary acquisition. No species among the mammals living today can be considered an ancestor of the bats, not even those that have become specialized as "gliders." In what stages and through what developmental steps the bats evolved into active fliers is still unexplained today. We must go quite a ways back in the fossil record to learn something about what the ancestors could have looked like. As fossil finds confirm, for example from the oil-shale pit of Messel near Darmstadt, Germany, bats lived 50 million years ago in the same form in which they are still found today. Studies of the structure of the inner ear and larynx of fossil bats show that they navigated at that time with the aid of ultrasound. Thus the bats are a very old, yet other placental mammals. It is therefore highly likely that bats evolved from primitive arboreal insectivores. The German name *fledermaus* (fluttering mouse) is thus misleading. However, it would hardly be possible in the long run to replace names that are rooted so deeply in the spoken

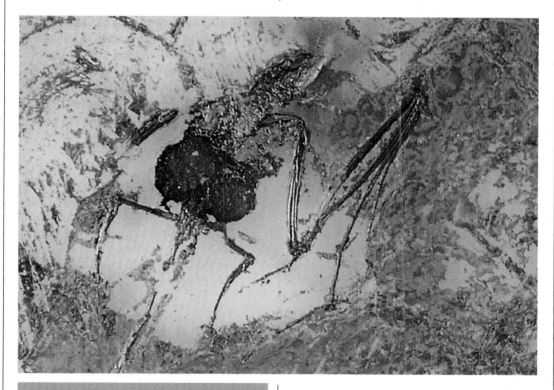

Fossil bat (*Archaeonycteris trigonodon*) from the Messel oil-shale pit near Darmstadt, Germany. The fossils of Messel are dated at 50 million years. This find is kept in the Senckenberg Museum in Frankfurt, Germany.

highly specialized, mammalian order.

We regard the insectivores, such as hedgehogs or shrews, as the most primitive placental mammals. Their ancestors can be considered as the ancestors of

language and literature with technically more correct ones, such as "hand flyer" or "fluttering animal."

Apart from the ability to fly and adaptations to new living spaces or new food sources that were acquired later, many bat species also exhibit quite primitive characters, such as in the structure of the skull, the shape of the dentition, the digestive system, or the degree of development of the brain. Of particular significance for the Chiroptera was the development of the ultrasound navigation system.

THEY FLY WITH THEIR HANDS

Bats were able to conquer the air as a result of the transformation of their arms and hands into organs of flight. The elongated forearm of bats consists only of the robust radius; the ulna has been lost. The second to fifth metacarpals and the fingers that articulate with them are also considerably elongated. The second finger consists of only one joint, the third of three joints, and the fourth and fifth of two joints. The cartilaginous tips of the fingers terminate in the shape of a "T" at the margins of the wing membranes. The thumb has retained its normal form and has remained short. It ends in a sharp claw, which the bats can use for climbing and hanging. In most flying foxes the second finger also has a small claw, which, however, is largely nonfunctional.

The wing membrane (patagium) stretches out from the sides of the body to between the fingers; the tail membrane (uropatagium) extends between the hind legs to the tail.

In most European bats, the tail is almost completely included in the membrane. An exception is the European Free-tailed Bat, in which most of the tail is free beyond the membrane. In comparison to the common bats, the horseshoe bats have a shorter tail, which they fold over the back when at rest. Common bats, in contrast, fold the tail in toward the belly, while the Free-tailed Bat usually keeps it extended.

Additionally, the margin of the tail membrane is stiffened and supported by a bony spur attached to the ankle joint. Depending on the flight maneuver and tension on the tail membrane, the spur lies either at the side of the foot or is splayed out from it. In certain genera, such as *Nyctalus*, there is also a stiff lobe of skin (epiblema) on the spur, which is supported in the middle by a T-shaped piece of cartilage (steg) extending from the spur.

The hind legs of the bat have several functions. They are usually included in the membrane down to the foot. With the aid of the legs the tail membrane is spread out or folded up in flight. The claws of the hind feet serve for hanging when the bat is at rest. As a result of the rotation upward and to the outside of the leg at the knee joint, the foot with its claws points to the rear and not, as in other mammals, to the front. This makes it possible for the bat to hang on walls. A special locking mechanism ensures that the claws do not release their hold, even without muscular tension, which is why even dead bats remain hanging in place.

Common bats and the free-tailed bats fold their wings together at rest so that the membranes are barely visible. The length of the wing is shortened in the majority of common bats by bending the third to fifth fingers 180 degrees at the basal joint and laying them against the

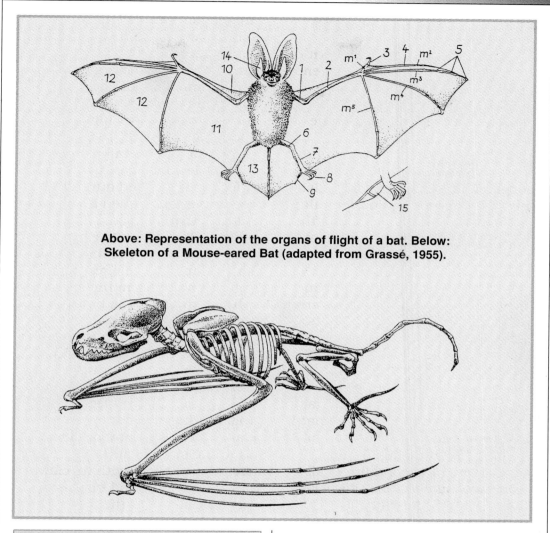

Above: Representation of the organs of flight of a bat. Below:
Skeleton of a Mouse-eared Bat (adapted from Grassé, 1955).

Components of the skeleton and wings:
1: humerus, 2: radius, 3: thumb, 4:
metacarpals (m1, m2, m3, m4, m5), 5:
phalanges, 6: femur, 7: tibia, 8: hind foot,
9: spur (without epiblema), 10: anterior
wing membrane (propatagium), 11: arm
wing membrane (plagiopatagium), 12:
finger wing membrane (chiropatagium or
dactylopatagium), 13: tail wing membrane (uropatagium), 14: ear cover (tragus), 15: spur with epiblema and steg.

metacarpals. Supported on the
wrist and the thumb, as well as on
the backward pointing feet,
common bats can lift the body
from the ground and run rapidly,
even jump or move sideways and
backwards. When climbing head
first, the thumb claws are used
alternately. Common bats can also
climb backward, and, with the aid
of the hind feet, hang head up.
They can even be observed
climbing in narrow cracks by
bracing themselves against the
sides of the opening.

Horseshoe bats are quite
helpless on the ground. They
cannot lift their body, only pull
themselves forward with their

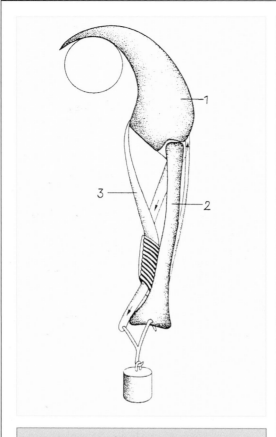

Representation of how the weight of the bat causes the claws of the hind foot to curve with the aid of a tendon. This results in a "passive" hanging of the bat on a ledge or branch in the resting phase (from Schaffer, 1905). 1: claw, 2: metatarsals, 3: tendon.

thumb claws. Therefore, we always find horseshoe bats hanging, never perched or running on a substrate. When they hang freely, on the other hand, they can move over a rough surface with small steps. They also always require an open entrance and exit to their roost. In the resting position, Greater and Lesser Horseshoe Bats bend the basal joint of the finger slightly and wrap it within the membrane located between the fingers as if in

a cloak. The Mediterranean Horseshoe Bat, in contrast, bends the fingers at the joint between the first and second phalanx 90 to 180 degrees and folds the membrane together in this area. This action does not enclose it completely.

Bats not only fly, climb, run, and jump, they also swim. Naturally, they do not do so willingly. However, if a bat does happen to fall accidentally into the water, it spreads its wings to swim. They often are able to take off from the water's surface.

The shape of the wing allows us to make inferences about the type of flight and the flying ability of a bat. The fast-flying species can readily be distinguished from the broad-winged, slow-flying species by their long, narrow wings. In flight, the wings carry out a type of rotating movement. The tail membrane helps in turning and in braking before landing. The thorax of the bat is very robust in comparison to the narrow pelvis. To improve the points of attachment for the flight muscles, the breast-bone, as in birds, has a keel.

Flight expends a great deal of energy, and the flight muscles, in particular, require a high oxygen supply. To meet this high energy requirement, which is about four times as high in flight as it is in the resting phase, great demands are made on the circulatory and respiratory organs. As a result, the heart rate and respiration frequency are greatly elevated in flight. The heart exhibits corresponding adaptations. It is

Above: The Barbastelle (*Barbastella barbastellus*) has relatively narrow wings and is a fast and agile flier.

Below: Flying Natterer's Bat (*Myotis nattereri*). Echolocation calls are sent out through the open mouth.

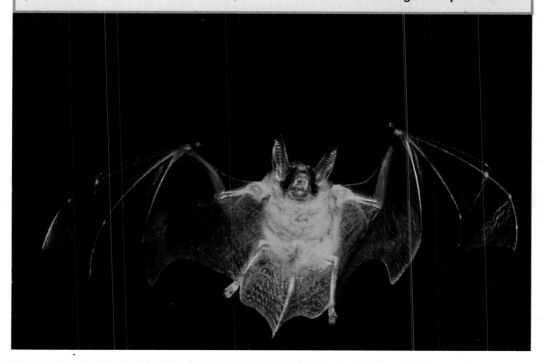

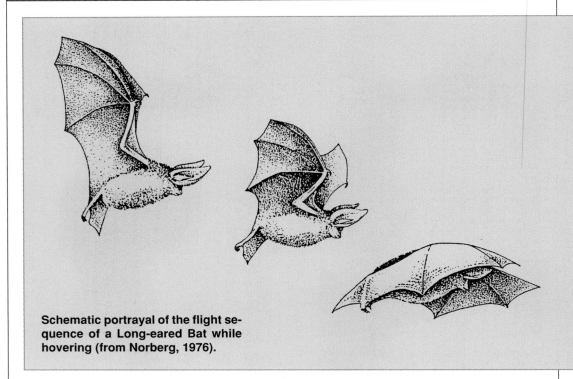

Schematic portrayal of the flight sequence of a Long-eared Bat while hovering (from Norberg, 1976).

about three times larger than the heart of other mammals of the same body size, and therefore can pump more blood through the body in the same amount of time. With the extra blood, the necessary oxygen is transported to the muscles. Bats have a further adaptation; whereas the oxygen uptake capacity is only 18% by volume in many mammals, it is 27% in bats.

Physical exertion produces heat, which causes the body temperature of the bat to rise in flight. Because bats can neither sweat like humans nor pant like dogs, an efficient cooling system prevents a dangerous rise in their body temperature: When the bat's body temperature rises, a special system of valves in the thin blood vessels in the wing membranes enlarge, and the now increased

flow of blood is cooled by the cold stream of air that constantly passes over the wings.

The wing membrane is a double membrane and includes, besides the blood vessels, nerves and small bundles of muscles. The muscles serve mainly to brace the wings so as to keep them from fluttering in the stream of air. The wing membranes, which are bare except for a part of the tail membrane, look delicate, but are very flexible and tough because of the elastic fibers they contain. They exhibit a tendency to heal well following injuries.

During embryonic development, the wing membranes arise from folds of skin on the sides of the body. The longitudinal growth of the wing bones begins later. The wings of newborn bats are still quite undeveloped. The hands of

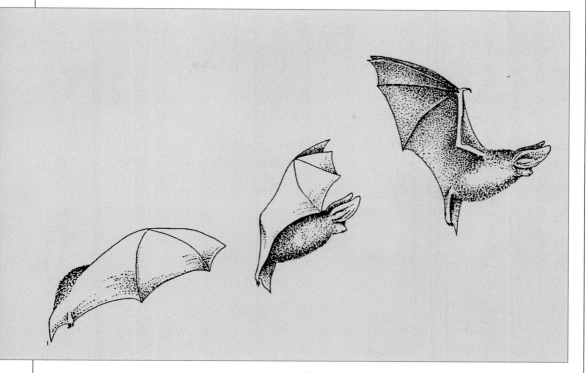

the young bats do not attain their ultimate proportions until they complete their growth outside of the mother's body.

Besides the wing membranes, the observer is impressed most by the face of the bat. When larger bat species threaten with an open mouth, they demand respect. The teeth resemble those of insectivores, but seem more like those of carnivores because of the large canine teeth. Noctules, Greater Mouse-eared Bats, and other large species can effortlessly bite through the hard chitinous exoskeleton of large beetles. The number of teeth in the European species varies from 32 (genus *Tadarida*) to 38 (genus *Myotis*).

Very conspicuous features in the horseshoe bat family are the bizarre skin formations in the nasal region. All common bats and the Free-tailed Bat lack these.

The ears of the bat are of variable shape and size. Their length varies from 9 to 13 millimeters in the common Pipistrelle up to a maximum of 42 millimeters in the Common Long-eared Bat. All common bats have an additional prominence (tragus) in the external ear, which the horseshoe bats lack. The variable shape and size of the ear and the tragus can also be used effectively to identify a number of bat species.

Bats have comparatively small eyes, which appear dark (or black) in all species. Comparatively, the long-eared bats have the largest eyes, whereas the Barbastelle has very small eyes. The eyelids lack eyelashes. When the eye is closed, the narrow crack between the lids is barely visible. A few fairly stiff sensory hairs are located in the mouth region of common bats and horseshoe bats, as well as on the upper part of the nasal region of the latter. Skin glands in the mouth region deliver an oily secretion for grooming the wing membranes. This apparently also contains pheromones, which send signals to conspecifics.

The fur of the bat, in contrast to

In many bats the hairs are two-colored. In all species the underside is lighter in color than the fur of the back; juvenile bats tend to be darker and duller in color than adults. In several species the fur coloration also changes somewhat following the autumn molt. (The variable coat coloration is best shown by the color photographs of the

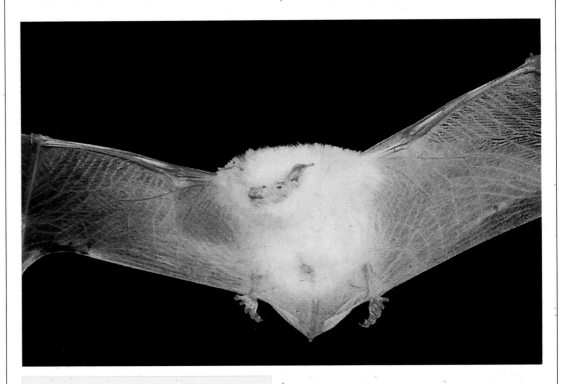

Albino female of Daubenton's Bat (_Myotis daubentoni_). Because of the absence of pigmentation, the blood vessels in the wing membranes are clearly visible.

individual species.) On the hind feet, most bats have more or less distinct, stiff, bristle-like hairs. On the free edge of the tail membrane, Natterer's Bat has a border of stiff hairs that are curved like hooks.

In a resting bat, the sex cannot be determined by appearance alone. Differences in coloration that, for example, occur in most birds do not exist in the European species. The difference

the majority of other mammals, contains no woolly hair; thus, it consists of only a single type of hair. Bat hairs have such typical structures that several genera or species can be identified by the hair structure alone.

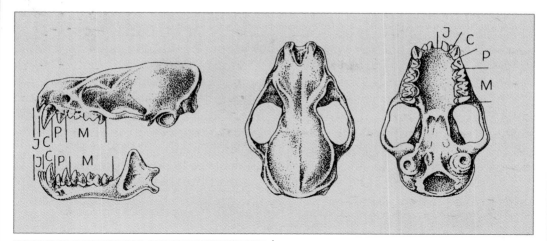

Skull of the Serotine (*Eptesicus serotinus*) in lateral, dorsal, and basal views. J: incisors, C: canines, P: premolars, M: molars). The teeth are well developed, pointed, and some have four cusps. They resemble the dentition of a predator. The pointed canine teeth seize the prey, and the sharp molars grind the chitinous exoskeleton.

in size between females and males (females as a rule are somewhat larger than males) cannot be distinguished without precise

Differences in the surface structure of the hairs of various bat species. a: Lesser Horseshoe Bat, b: Natterer's Bat, c: Nathusius's Pipistrelle, d: Barbastelle, e: Noctule, f: European Free-tailed Bat. Shown in each case are sections taken from the middle of the hairs (from Tupinier, 1973).

measurement. When we examine the genital region in the male, the penis is readily visible. Furthermore, during the mating season the testes and epididymis of certain species protrude and are clearly visible, indicating the bats' readiness to breed. The females have a pair of mammary glands, which are located in the region of the armpit. (An exception among the European bats is the particolored bat, which has two pairs of mammary glands.) The teats are clearly visible in nursing females. Horseshoe bats also have a pair of so-called "false" teats in the lower belly region, to which the young can hold firmly with their mouth.

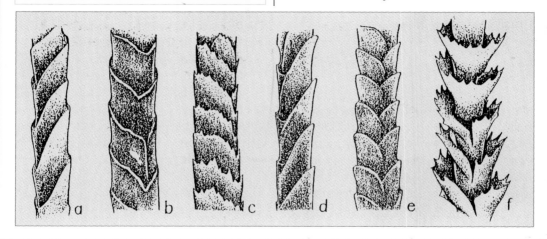

WHERE DO BATS LIVE?

All European bats require places to roost that offer them protection from the unfavorable effects of weather (cold, rain, drafts, etc.) and from enemies or constant disturbance. Because bats do not build nests, they are dependent on existing hiding places.

According to their biological function, we can distinguish the following types of roosts: winter roosts, day and temporary roosts, maternity roosts, and mating roosts. The not quite precise term "summer roost" has become established for the latter three types of roosts.

The bats hibernate in their winter roosts. Various species often gather in large numbers in a single roost. The largest known current winter roost in Europe is the one in the bat reserve "Nietoperek" located at Miedzyrzecz in western Poland. Here, well over 10,000 bats of 12 species hibernate about 50 meters below ground level in an immense system of concrete bunkers and passages from World War II.

When they awaken from hibernation, the bats, usually separated by species, seek out roosts in which they often stay for only a few days, but sometimes several weeks. In these day or temporary roosts, we sometimes find only individual bats, but occasionally small groups as well. They visit their hunting territory from here each night. These temporary roosts are occupied by the bats on their migrations from the winter roost to the places where they stay in the summer. Maternity roosts are accommodations that are shared by a more or less large number of female bats for a period of several months. The females give birth to and rear their young in these maternity roosts. The males of many species live alone in their day roosts during this time. Some species, such as the Parti-colored Bat and Noctule, form fairly large male groups. After the maternity roosts disband, the males and females come together to mate. As a rule, the mating roosts do not differ from the day roosts. With the bats living in northern and central Europe, the four types of roosts are usually spatially separated from one another. Migratory species cover large distances between the winter roost and the summer roost. On the other hand, in the milder climate of southern Europe the same bat species can live throughout the year in a single roost, for example in a cave. In sedentary species the different roosts are located close together, even in central Europe. The Common Pipistrelle, for example, hibernates in a church behind memorial slabs or in cracks in walls, and then occupies day roosts in the church nave or cracks in the outside walls. Later, a crack accessible from the outside behind the boards of the church steeple can serve as a

Facing page: Roost of the Noctule (*Nyctalus noctula*) in a plane tree. Clearly visible is the urine stripe running out of the hole.

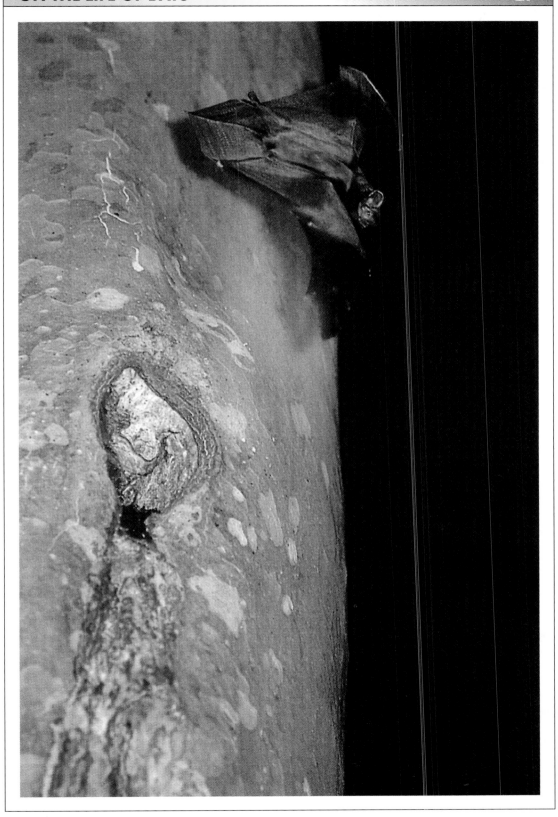

maternity roost. Mating then takes place again in the day roost.

A further division of the bat roosts can be made according to the biotopes and localities they inhabit, particularly in the summer. House bats (anthrophilic bats) are more or less tied to human settlements. Typical species of house bats in central Europe are the Serotine, the Mouse-eared Bat, and the Gray Long-eared Bat, and, with restrictions, the Pond Bat, the Whiskered Bat, the Lesser Horseshoe Bat, and several other species. Depending on climatic conditions, the same species can live as a house bat in the northern part of its range and as a cave bat in the south. The Lesser Horseshoe Bat, for example, inhabits warm attics or heated basements in the north in the summer, whereas it is a cave dweller in the south. As warmth-loving animals, bats could only extend their range farther to the north by seeking out buildings with more favorable microclimates. Natural caves are too cold in these regions, but the "artificial caves" of humans offer favorable living conditions.

Among the species living in southern Europe primarily as cave-dwelling or cliff-dwelling (lithophilic) bats are: all horseshoe bats, Schreiber's Bat, and the Lesser Mouse-eared bat. Forest-dwelling or tree-dwelling (phytophilic) bats, on the other hand, are bound to the forest in summer throughout their entire range. Even in winter some of them hibernate in tree holes.

Typical representatives of phytophilic bats are three species of Noctule, Nathusius's Pipistrelle, and Bechstein's Bat. When we examine the different bat roosts and the hanging places more closely, an inexhaustible wealth of roosting places is revealed.

The winter roosts are relatively uniform. They must meet the following requirements: The temperature must not, at least in some parts of the roost, fall below 0°C, the relative humidity must be high (up to 100%), and there must not be any draft. The walls must be rough, in order to provide the bats places from which to hang. Also favorable are additional shelters, such as cracks in walls and small niches or hollows. Very badly soiled roosts, fresh lime or mortar, and soot deposits are avoided by bats. Frequent disturbances, such as the use of a cellar as a storage area, have an unfavorable effect, but do not absolutely exclude the space from being used as a roost. In mountainous regions, natural caves and old mine shafts are ideal winter roosts. In the lowlands, underground foundations, old beer and wine cellars, former ice cellars, vaults, old bunkers, and deep basements in homes are used. Tree bats overwinter in tree holes with wall thicknesses of more than 10 centimeters, but also in deep cracks in cliffs and walls, or smaller spaces in thick-walled buildings. Serotines also occasionally hibernate in attics in narrow cracks in the rafters.

In their winter roosts, certain species hang freely exclusively (horseshoe bats) or predominantly (mouse-eared bats) from the ceiling. Other species, on the other hand, hang from the walls or seek out cracks

There exists a host of possibilities for day or temporary roosts. House bats can hide in the roof truss, under roof joists, in holes in beams, or in the ridge. On the outside of the house, the possible roosts include cracks

Old fortifications often serve as winter roosts for bats.

in the walls, sometimes even in the rubble on the bottom of caves. Even open holes or bored holes are used as roosts. Furthermore, hibernating bats have also been found behind pictures in churches, crumbled plaster, peeling layers of rust on iron girders, or on roots growing through the ceiling of the roost.

behind shutters, on siding, between beams and masonry, behind the flashing of chimneys, and behind Venetian blinds. When bats enter a house, they often hide in the folds of drapes. When you look for bats in an attic, you should not look up, but rather should first look for droppings on the floor. It is then easy to find the roosts located overhead.

Forest bats make their day

roosts in abandoned woodpecker holes, rooted holes in old trees, cracks in trees, in spaces behind loose bark, and in wood piles. They also readily accept bird houses and special bat boxes.

Cave and cliff bats seek out not-too-cold natural caves, tunnels, or crevices. However, they also make use of a whole series of atypical day roosts. For

Maternity roosts usually are roomier than day roosts, because they must provide enough space for a group of females and their young. For their maternity roosts, house bats seek out the warmest locations inside the attics of houses, churches, or other buildings, often in the roof ridge or in the church steeple. Unsuitable are ruins with half-

Places where bats roost in buildings and trees.

example, they can occupy the nest holes of bank swallows, cracks under bridges, hollow concrete poles of street lights, ventilation shafts, and gaps between the wall sections of apartment buildings.

destroyed roofs, because they are cold and drafty and the rain leaks in. The roofs of buildings with bat roosts must be intact. Naturally, an entrance hole, either for flying or crawling in, must be present.

House bats, which prefer narrow cracks, establish their maternity roosts behind shutters and boards on the outside of the

Pile of droppings with dead young. With the Greater Mouse-eared Bat (*Myotis myotis*), in the northern part of the range a majority of the only several-day-old young die in periods of bad weather.

building. Common Pipistrelles occupy suitable cracks in newly constructed private homes and bungalows relatively quickly. The Lesser Horseshoe Bat and Mouse-eared Bats are known to occupy maternity roosts in warm underground spaces, for example underground heating shafts or cellars.

The maternity roosts of forest or tree bats are located in roomy tree holes or in bat boxes. If a tree hole is occupied for a fairly long time by a fairly large number of bats, it can frequently be observed that traces of urine are present as dark stripes at the hole entrance and that droppings collect under the "cave." (Bats in tree holes also reveal their

presence to experts by their calls.)

The maternity roosts of cave bats are always located in the warmest part of the cave. They can contain more than a thousand individuals that, depending on the species, hang close together on the ceiling, frequently also unite in fairly large groups, or fill niches in the cave.

What is also true of all bat species is that they have several alternate roosts, which they change from time to time. Why this alternation takes place is largely unknown. The worsening of the microclimate certainly is just as significant as outside disturbances, mass infestation of parasites, or a change in the hunting territory because of food shortages. As a rule, however, bats return to their roosts year after year, and maternity and winter roosts in particular are inhabited over many decades.

HUNTING PREY WITH ECHOLOCATION

All European bats feed primarily on insects, although they exhibit differences in the selection of the insect species they prefer, their hunting territories, and their hunting strategies. These differences make it possible for several bat species to live in the same biotope without constant competition. Examples of this are the Noctule and Nathusius's Pipistrelle in the forest, or the Mouse-eared Bat, the Common Pipistrelle, and the Gray Long-eared Bat in human settlements. Our knowledge of these slight ecological differences is still very fragmentary, so that a potentially rewarding area of research presents itself. The more we know about the habits of a bat species, the easier it is for us to offer it systematic protection.

When we observe a bat roost on a summer's evening, we notice that with great precision the bats always leave the roost at the same time to hunt. The beginning of the emergence is announced by a considerable unrest in the roost, which is revealed in the Common Pipistrelle, for example, by a soft high-pitched whispering. At this time the bats stay near the entrance to the roost. It is also possible to observe that the bats first groom themselves intensively and stretch their wings. Once the first bat has taken flight, the others follow at short intervals. The emergence from a maternity roost of Common Pipistrelles, numbering about 100 individuals, can take as long as an hour. In cool weather bats set out later to hunt, and in a strong wind or rain they may not emerge at all. The time the sun sets and the amount of light in the vicinity of the roost have a decisive influence on the timing of the beginning of the hunting flight.

There are differences between the species in regard to the time they emerge from the roost. These are even reflected in names for bats in the old literature. The Noctule, for example, was designated as the "early-flying bat," the Serotine, on the other hand, as the "late-flying bat."

Particularly in the spring and fall, we can observe the native bats hunting even during the day. Often these are Noctules, which already hunt in early afternoon, but Common Pipistrelles and other species are occasionally active during the day as well. Apparently, the poorer food supply forces the bats to adopt this hunting behavior. In some observations of day-flying Noctules, however, it can also be a question of migratory activity.

The duration of the hunting flight varies from species to species. It is strongly dependent on the available food supply. Before bats start to hunt they usually seek out ponds, streams, and other sources of drinking water. In order to drink they repeatedly fly so close to the water's surface that they can briefly submerge their mouth. Whether they drink in the wild while perching or hanging, as they do in captivity, is not known for certain, but it certainly seems possible.

Bats have regular hunting territories, the location and size of which depends on the species, the season, and the food supply. Nyholm established a hunting territory size of 240 square meters for the Whiskered Bat in a forest biotope; on the other hand, the territory size was 420 square

swarms of insects are present, however, such as around flowering trees, several bats sometimes hunt there together. The hunting territories themselves are very fluid and are determined by both the current food supply and the preferred living space of the bat. Certain forest bats prefer

Drinking Bechstein's Bat (*Myotis bechsteini*).

meters for Daubenton's Bat. The bats often maintain fixed flight paths in these territories. They fly circuit after circuit until they have "cleaned out" the insects on these paths, after which they switch to another flight path. Where

to hunt low over forest meadows, forest paths or clearings, along forest margins, and in firebreaks. Others, such as the Noctule, usually hunt above the treetops or high above the surface of lakes and ponds. Daubenton's Bat exploits the abundant insect life in the region of the riparian zones of lakes and ponds, where it hunts just above the water's surface. Brandt's Bat and the Pond Bat also prefer to hunt in these places. While hunting, bats often follow the course of a stream or river. House bats usually hunt in towns, on farms, in gardens, over garbage dumps, or around street lights, which attract insects to their bright light. For some species, such as the Common Pipistrelle, there are ample places to hunt, even in cities.

Bats with long, narrow wings, such as the three noctule species, the European Free-tailed Bat, and Schreiber's Bat, prefer the open air space. They fly very fast and maneuver poorly in confined spaces. The noctule reaches a maximum flight speed of 50 kilometers per hour, Schreiber's Bat reaches 70 kph, and a relative of the European Free-tailed Bat, the Guano Bat (*Tadarida brasiliensis*), is even said to attain 105 kph.

Smaller species, such as the Barbastelle and Daubenton's Bat, also fly quite fast, but are more agile in confined spaces. All fast-flying species have very short ears, whereas some slow-flying bat species have very large ears.

The Serotine and Mouse-eared Bats, which hunt in slow flight in the open air space, have long, broad wings. Short, broad wings with a relatively large wing surface, on the other hand, are best suited for hunting in dense vegetation or in landscapes with numerous obstacles. These species (for example the Common Long-eared Bat) can search the countryside slowly at low altitude

and can even hover in place to glean prey animals from twigs and leaves. The horseshoe bats are particularly agile fliers, though the long-eared bats, Bechstein's bat, and Natterer's Bat are hardly inferior to them in this respect.

Bats locate insects with ultrasound, then chase them down and catch them directly in the mouth or by using their arm membranes like a butterfly net. From the arm membrane they are usually taken up directly into the mouth. Sometimes the prey ends up in the tail membrane, which is curved toward the belly like a pocket, from which it cannot escape. From there it is then consumed in flight. The long, sharp canine teeth of the insectivorous bats penetrate the prey and hold it tight. Finally, the bat grinds the sometimes very hard chitinous exoskeleton with its broad, multi-cusped molars.

Most bats devour their prey in flight. With larger insects they also return repeatedly to a familiar "parking space" without using echolocation. Slow-flying species, such as the horseshoe bats and the long-eared bats, go to a permanent feeding roost with their prey animals. They hang there and devour the body of the prey. While feeding, they drop the wings and legs of butterflies and moths or other insects to the ground. By means of such accumulations of wings in the roosting areas it is possible to confirm the presence of bats. The zoologist Kolb describes the Mouse-eared Bat as also occasionally taking flying or flightless beetles from the ground. They are first located with their sense of hearing and then tracked using their well-developed sense of smell. Smell definitely plays a role in prey capture in all insectivorous bats, because it allows them to sort out inedible and bad-tasting insects. Whether visual orientation also plays a role in prey capture in certain species is not precisely known. The Common Long-eared Bat does not react to a motionless butterfly in the laboratory, but seizes it immediately when it starts to

Because all species are strictly protected, the study of the stomach contents of killed bats is out of the question. Therefore, we must turn to the analysis of food remains, which is relatively easy with some species because of the butterfly and moth wings at the feeding roosts. The method of stool analysis, on the other hand, in which prey animals can be identified from the smallest remnants of chitin, requires considerable experience and patience. This analysis shows, however, that, besides butterflies,

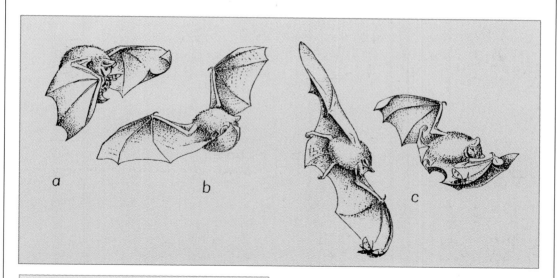

a   b   c

Capturing the prey. As a rule, insects are seized directly with the mouth in flight (a). Larger prey animals are often captured in a pouch formed by the tail membrane and are taken from there into the mouth (b). If the prey animal tries to escape, it is "caught" with the wings and directed toward the head (c).

move. Maybe optical stimuli also play a role, along with acoustic detection.

The dietary spectrum of many bats is still largely unknown.

moths, beetles, mosquitoes, and flies, bats also prey on dragonflies, crickets, grasshoppers, scorpion flies, and spiders.

In the course of their evolution, some of the prey animals have developed defensive mechanisms against bats. For example, they can detect certain frequencies of their ultrasound calls. Most of these prey animals use this information to fly away from the ultrasound beam, while others simply drop to the ground. Some moths, however, use ultrasound calls that they produce themselves to signal the bats that they taste bad. They are then avoided by the bats after their first unpleasant experience.

The daily food requirement of bats is between a fourth and a third of their body weight. According to studies by the Russian zoologist Kurskow, the maximum daily food intake in the wild was 38% in the Noctule, 31.3% in the Parti-colored Bat,

29.5% in the Common Pipistrelle, and 28% in the Barbastelle. The calculation that a single Daubenton's Bat can destroy 60,000 mosquitoes from May 15th to October 15th is certainly somewhat theoretical, but it does a very good job of illustrating the dimension of the food requirement. We can also get an idea of the mass of the animals consumed when we see the sometimes meter-high mountains of droppings in bat maternity roosts that have been occupied for many years.

Many bat species consume large numbers of insects that are pests of agriculture and forestry. For example, the prey insects of the Common Long-eared Bat

Feeding roost of the Common Long-eared Bat (*Plecotus auritus*). Besides various noctuid species, butterflies such as the Tortoiseshell Butterfly and the Peacock Butterfly are also preyed on.

include the Seed Moth (*Agrotis segetis*), the Vegetable Moth (*Polia oleracea*), and the Oak Roller Moth (*Tortrix viridana*). This, in the best sense of the word, biological pest control by bats unfortunately has become less significant because of the substantial decline in bat populations.

pitched humming of a large bumblebee. The Noctule has the loudest voice. Its shrill, metallic-sounding calls are audible over distances of up to 50 meters or so. These calls, as do a portion of the ultrasound calls, have a social function and serve as communication between the animals. There are contact calls

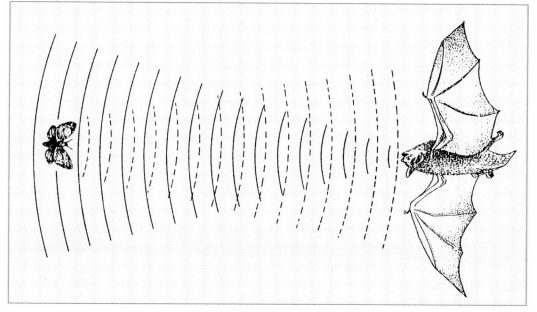

Location of the prey with the aid of the reflection of the transmitted sound waves by the prey animal.

ECHOLOCATION—THE SIXTH SENSE OF BATS

Bats produce both audible calls with a frequency below 20 kHz and sounds in the ultrasound range above it, and therefore are inaccessible to the human sense of hearing. The audible calls vary greatly. They can sound like a shrill scolding, a soft peeping, or a chirping, or call to mind the low-

between mother and youngster, calls that are emitted during defensive or aggressive behavior, cries of pain, and mating calls, with which the males attract ready-to-mate females.

The ultrasound calls of bats were discovered and their significance recognized very late.

Since the eighteenth century, researchers had tried to solve the puzzle of the navigation of bats, but it took until 1938 for the American Griffen to come up with the explanation. He determined that bats transmit ultrasound calls and receive the returning echoes with their ears, thereby perceiving their environment. Humans and most mammals orient themselves with their eyes; we produce an optical picture of our environment and store it in our memory. Bats, on the other hand, obtain a "sound picture" from their environment that likewise is precise and "in color." Instead of seeing, they hear the entrance to their roost, the ledge from which they can hang, and the prey animals that cross their flight path. They store the sound picture in their memory and can automatically avoid an obstacle in a familiar environment while "flying blind." A similarly efficient orientation with the aid of ultrasound is also found in such mammals as whales and dolphins. Cave-dwelling birds, including the South American Oilbird (*Steatornis caripensis*), and cave swiftlets of the genus *Collocalia*, also orient themselves using the echolocation principle. The frequency range of the orientation calls ranges from 20 to 215 kHz. With an increasing frequency of oscillation per time interval, the wavelength becomes increasingly shorter. This is advantageous, precisely because the short waves are best suited for producing useful echoes from small obstacles. When we make the signal visible, which is possible today even in the wild with an ultrasound detector (bat detector), it turns out that many species produce a specific frequency picture (sonogram) and transmit a species-typical

frequency. With common bats, the ordered frequency sequence falls off steeply at the end of the call. The call, which includes only 50 oscillations, often runs through a frequency range of 100 kHz down to 40 kHz. If we were able to hear such a call, then it would not be a pure tone, but a burst of sound. Because of the typical frequency modulation of their ultrasound calls (FM signals), these bats are also called "frequency modulating bats." In recent years it has repeatedly been determined that the diversity in the range of the echolocation calls is much greater than initially suspected. It turns out that some bat families use a completely different system of pictorial hearing and that some species use several systems.

It was determined, for example, that the horseshoe bats do not transmit bursts, but rather pure tones lasting up to 150 milliseconds, with only a short frequency-modulated end portion. At the same time, the frequency is very typical for the individual species. The Greater Horseshoe Bat transmits at 83 kHz, whereas the Lesser Horseshoe Bat uses the transmission frequency of 107 kHz. Because of the constant frequency of the transmitted ultrasound, with only a very short frequency-modulated end section (CF/FM signals), the horseshoe bats are also called "constant-frequency bats."

The native common bats transmit their echolocation calls out of their open mouth. The two long-eared bats are an exception, because they can also transmit

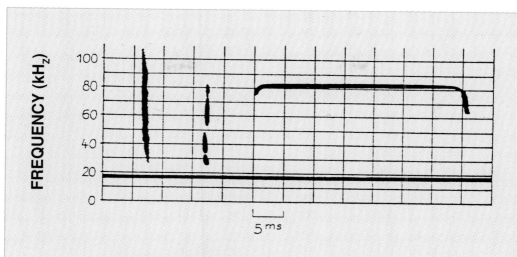

DURATION (SEC)

Ultrasound echolocation calls of various bats. The common bats (Mouse-eared Bat, Long-eared Bat) emit short calls that go through a large range of frequency in a few milliseconds (FM calls). The horseshoe bats echolocate with long-lasting calls in a very narrow frequency band (CF calls, pure tones). The upper threshold of sounds audible to humans is at 18 kHz.

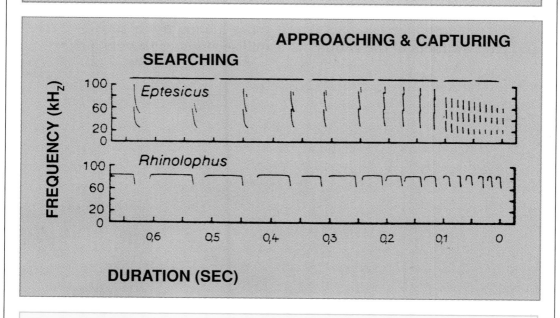

Sonograms of the echolocation calls of common bats and horseshoe bats in various hunting phases. Once the prey has been located, the pauses between the calls become increasingly shorter in the approach and capture phases. In the horseshoe bats the duration of the call is also shortened.

ultrasound through their nostrils with the mouth closed. The horseshoe bats also transmit ultrasound with a closed mouth, with the flexible parts of the nose leaf surrounding the nostrils focusing the sound. This beam of sound is swung back and forth like a flashlight using swaying movements of the head. The returning echoes are collected by the ears, which can move independently of one another.

In the common bats with their FM signals, it is typical that the individual sounds are of only very short duration, in order to prevent the transmitted call and the echo from overlapping. Because ultrasound spreads in air at a velocity of 331 meters per second, an FM impulse may be maintained for only one-half to one millisecond if the echo reflected by a prey animal about 50 centimeters away is to be heard without distortion.

The frequency modulating species determine the distance to the prey animal from the time difference between the transmission of the signal and the arrival of the echo. On the other hand, the direction is determined from the analysis of the time difference between the arrival of the echo at the right and the left ears. Also important are the intensity and the frequency of the signal. The echolocation system of horseshoe bats is not based on time-differential orientation, because the echo overlaps with the transmitted sound because of the long impulse duration of the CF/FM signals. The horseshoe

bats have a filter in their ears that is tuned precisely to the species-specific frequency. As a result, the Greater Horseshoe Bat hears its species-specific echo of 83 kHz considerably better than the frequencies above or below it. Because the echolocating horseshoe bat flies toward the echo returning from a prey animal or obstacle, the number of oscillations that it will hear per time interval will become increasingly greater and therefore increasingly higher in pitch. Through this effect, which is named after the physicist Doppler, the frequency of the echo would be higher pitched than that of the transmitted sound, and there would be the danger that the echo would not be heard at all. Yet horseshoe bats are able to balance out this effect by starting the transmitted call at a correspondingly lower pitch, so that the echoes have precisely the species-specific frequency when they are received. If the orientation call encounters an immovable obstacle, the echo returns as a pure tone. A flying prey animal, on the other hand, modulates the frequency of the echo with the rhythm of its wing-beats. By means of their orientation system, not only can horseshoe bats properly interpret the origin of the echo, they are even able to detect fine differences

Facing page: Just before taking off, the common bats echolocate through the open mouth. Shown here is an echolocating Brandt's Bat (*Myotis brandti*).

in the surface structure of obstacles.

A simple experiment reveals the sensitivity of this system. If you throw a pebble in front of a hunting bat circling in your garden, with a sudden turn it will fly immediately toward it. However, because it recognizes that it is not a prey animal, it turns away just before reaching it without attempting to capture it.

The performance of the echolocation system of the horseshoe bats is clearly better than that of the common bats. In experiments, horseshoe bats can still detect wires with a diameter of only 0.08 to 0.05 millimeters. On the other hand, common bats, such as the Common Long-eared Bat or the Lesser Mouse-eared Bat, can only detect wires down to a diameter of 0.2 millimeters. Therefore, horseshoe bats are caught less often in nets than are common bats. The range of their echolocation system also gives the horseshoe bats an advantage over common bats. They have a maximum range of 20 meters and can still detect a prey animal at a distance of 8 meters. Common bats, on the other hand, must approach to within about 2 meters of obstacles to be able to detect them.

There are differences between the species with regard to the loudness of the transmitted ultrasound signals. The Noctule, hunting in open air space, "screams" its echolocation call into the night with an intensity of sound of up to 100 decibels, which is comparable to the loudness of a jackhammer. On the other hand, the species that fly slowly and at low altitude between trees and bushes, such as Bechstein's Bat and long-eared bats, only "whisper." When they hunt in open air space, bats transmit fewer echolocation calls than they do in a countryside with numerous obstacles. When they approach the prey, the number of transmitted impulses is also increased greatly, at which time it can amount to up to 100 per second.

In their first days of life, young bats can produce only relatively undifferentiated, low-pitched sounds, which they use to make contact with their mother. The calls of young Mouse-eared Bats are at a frequency of 20 to 30 kHz up to an age of 18 days. Only then does the frequency increase slowly to 50 to 70 kHz.

As children learn to walk, young bats must learn to echolocate. In some species the juvenile first flies after its mother, which "tows it along" with sound. Juveniles of other species train alone in or just outside of the roost while the mother is away hunting.

The problem of echolocation in bats is far more complicated than can be presented here. A series of questions still require explanation. Still unsolved is the manner of long-distance orientation in bats. Whether they determine their direction by the position of the stars in the night sky or whether they orient themselves to the earth's magnetic field has yet to be studied. Long-distance orientation makes it possible for bats, even in regions that are unfamiliar to them, to return to their roost. Experiments in which Banded Common Pipistrelles were transported to different locations and released showed that they still found their way back to their roost from a distance of 50 to 60 kilometers.

Let us now turn from pictorial hearing through echolocation to the performance of the other senses:

Bats are able to see. In most species the eyes are small and the vision is not acute. They can detect differences in brightness and shapes, but they are unable to see in color.

Their senses of smell and taste are well developed. The sense of smell, for example, plays a large role in the recognition of mother and youngster. In certain species it also helps in locating prey. Prey animals that taste bad are avoided by bats after their first negative experience.

Sensitivity to temperature is very well developed and is of great significance for choosing a suitable winter roost.

The sense of touch is improved by the sensory hairs in the facial area and on the feet. Bats can detect slight currents of air particularly well, which is significant in the search for roosts.

HAREMS AND MATERNITY ROOSTS

Our knowledge of the reproductive biology of a number of European bat species is still very sketchy. The mating season begins after the maternity roosts disband in late July to late August. It includes the time of hibernation and extends into the spring. In the mating season, the testes and epididymis of the male, which otherwise are not outwardly visible, protrude clearly. In males of Nathusius's Pipistrelle, the appearance of the face also changes because of the swelling of either side of the ridge of the nose. Furthermore, when the bats open their mouth, yellowish ridges containing glands are visible at the inside corners of the mouth.

Permanent pair bonds are unknown in the European bat species. A male mates with several females. Apparently a female can also mate with different males. In a number of bat species (for example, the Noctule and Nathusius's Pipistrelle) the males are solitary during the mating season and establish a harem of about 2 to 10 females.

A male Nathusius's Pipistrelle, for example, occupies a bat box as a roost and stays in it for several days to weeks. The mating roost and the surrounding territory are defended against other males. Through mating calls given from inside the box or during flight in the breeding territory, the male tries to attract females and then chases them in flight. The males often retain their breeding territory for several years. On the other hand, females change territories, thus joining the harem of another male.

Copulating Daubenton's Bats (*Myotis daubentoni*) in the winter roost.

If copulation takes place in the winter roost, the female's role is at first passive. Observations of Daubenton's Bat showed that a male awakened from hibernation first begins to search for females. The male flies to its conspecifics hibernating individually or in groups, lands next to or on them, and begins to seek out a female, apparently by examination of its scent. The male behaves quite energetically during this inspection, and it can awaken a whole group of bats during its search. When it has found a female, the male embraces it from behind with its wings and holds it tightly. Copulation begins only when the female awakens slowly from its torpid state. At this time, however, it is not yet capable of flight. During copulation the bats occasionally produce shrill calls. The male occasionally bites the female's nape fur. The animals remain in the mating position for twenty minutes or longer, during which time it is possible that several copulations can take place.

In bats of the temperate climatic zones, the egg is not fertilized at the time of copulation. The sperm is stored and kept viable in the female's reproductive tract through the hibernation period and into spring. The maturation of the egg, fertilization, and the subsequent embryonic development are delayed until the female awakens from hibernation. This is a unique phenomenon among mammals.

Schreiber's Bat is an exception, achieving the same effect but in another way. In Schreiber's Bat the egg is fertilized immediately following copulation, but embryonic development—as in the Roe Deer (*Capreolus capreolus*)—stops at the blastocyst stage and does not continue again until the spring. In both cases this adaptation ensures that the young will not be born until the warm season. When female bats must be cared for in captivity, for example after the destruction of their winter roost, they should not be kept at warm temperatures because fertilization and embryonic development will occur prematurely. The young will then be born in the spring or winter. Because the precise time of fertilization can scarcely be established, it is difficult to determine the gestation period of bats. The data range from 45 to 70 days. Most European bat species give birth once a year to one youngster. In some species twin births are the rule, and three young have been observed in rare instances. This extremely low reproductive rate, in comparison to other mammals of the same size (a female house mouse [*Mus musculus*] can bring up to 90 young into the world in a single year), is balanced out by a high life expectancy and the early onset of sexual maturity. Bats can live up to 30 years. The average life expectancy, however, depending on the species, is considerably lower, about 2 -4 years. Migratory species, such as the Noctule and Nathusius's Pipistrelle, apparently are exposed to substantially more risks than are sedentary bats. They compensate for their low life expectancy through the mating of the females in the first year of life and through twin births. In the Common Pipistrelle, the females also take part in breeding in the first year, which is also the case in a lower percentage of Mouse-eared Bats and Lesser Horseshoe Bats. In the second year of life, all females and males participate in breeding. However, there are always a few females that do not give birth.

Starting in April, the females gather in maternity roosts, which, depending on the species, can include 10 to more than 1000 individuals. Males are rarely found in these maternity roosts. The female bats scarcely enter a torpid state during the day in the maternity roosts because the resulting lowering of the body temperature would lead to a delay in the embryonic development. If cold weather forces an extended torpid phase, the young are born later. The females give birth in the maternity roost over a period of several days to weeks.

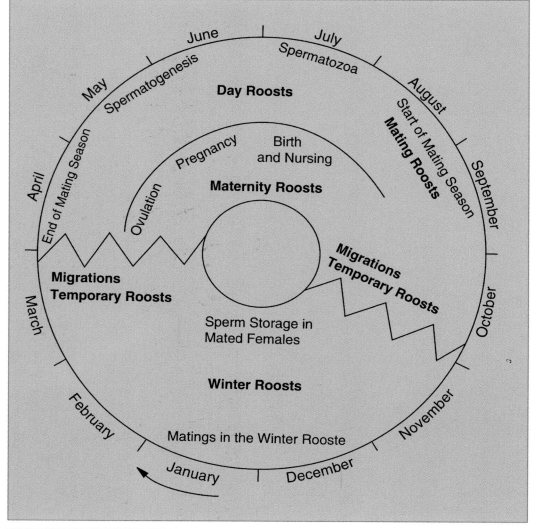

June

July

May   Spermatogenesis   Spermatozoa

April   End of Mating Season

Day Roosts

August
Start of Mating Season
Mating Roosts

September

Pregnancy   Birth
and Nursing

Ovulation

Maternity Roosts

March

**Migrations
Temporary Roosts**

**Migrations
Temporary Roosts**

October

Sperm Storage in
Mated Females

Winter Roosts

February   Matings in the Winter Rooste   November

January   December

Reproductive cycle of the European bats

The birthing process has been observed closely and described for various bat species, including the Mouse-eared Bat by Kolb and Nathusius's Pipistrelle by Heise. Birth occurs primarily during the day. The female separates herself somewhat from the others and assumes the birthing position. In this position, it clings to the substrate with all four limbs and perches either with the head down (Mouse-eared Bat) or in the otherwise atypical position with the head pointing up (Nathusius's

Facing page: Top: Portion of a maternity roost of Geoffroy's Bat (*Myotis emarginatus*) in an attic. In the lower center is a dark-colored youngster. Bottom:The newborn twins of the Common Pipistrelle (*Pipistrellus pipistrellus*) have immediately attached themselves to the mother's teats. They are almost naked, the ears are limp, and the eyes are still closed.

and Common Pipistrelles). Gebhardt also describes a horizontal birthing position for the Noctule with the back pointing downward. The hind legs

after birth by the pocket formed by the tail membrane, in which it is caught. The umbilical cord can also save the youngster from a fall by acting as a safety line.

Whiskered Bat (*Myotis mystacinus*) with nursing youngster. The mother protects it by forming the tail membrane into a pouch. Tiny mites are visible on the mother's wings.

are spread slightly, and the tail membrane is folded in toward the belly like a pocket. The young are usually born feet first, but they can also be born head first. The youngster is protected from falling

Newborns that have fallen usually are not picked up from the ground by the mother and eventually die. The birth weight reaches about one-third to one-fifth of the mother's weight in species that give birth to one

Facing page: Mother bats carry young bats in flight only when disturbed or when changing roosts. Here a Greater Mouse-eared Bat (*Myotis myotis*) with youngster.

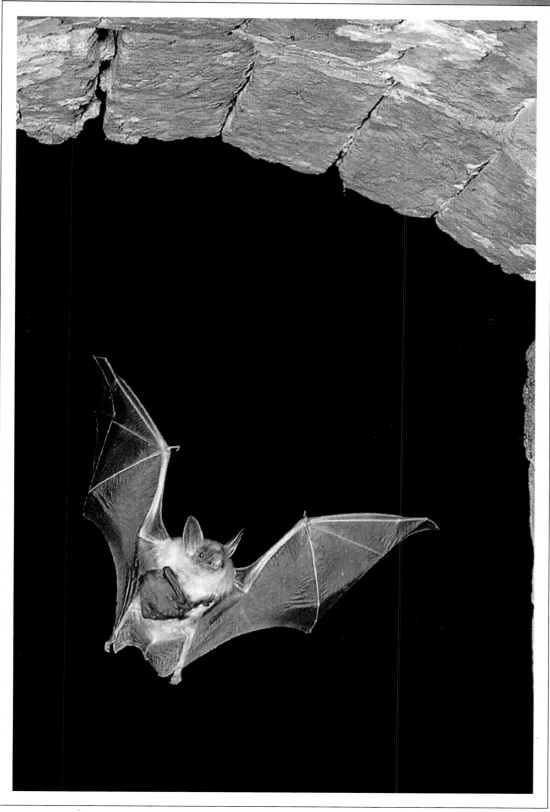

youngster. It is lower in species that give birth to two young. With the Common Pipistrelle, the pink, almost completely naked, and blind young are scarcely larger than bees. They weigh 1.3 to 1.5 grams.

Immediately after birth the young climb up the mother and suckle on a teat. The placenta is expelled later and is usually eaten by the mother. At the same time the mother bites off the umbilical cord on the youngster. The young are licked intensively immediately after birth. While being licked the young give high-pitched, twittering calls, which subsequently serve as contact calls, making it possible for the mother and youngster to recognize each other. Until they become independent, the young are nourished with mother's milk. The mothers do not bring the young along on the hunting flight at first, but rather loosen them from the teat with gentle nudges of the head and leave them alone in the roost. The young then often hang close together in fairly large groups (social thermoregulation). Nevertheless, when she returns from the hunting flight, each mother recognizes her own youngster, which is the only one she nurses. The young also recognize their mothers. This recognition is made possible through contact calls as well as through specific scents. Foreign youngsters are turned away. When disturbed, the mothers also take their young along in flight. In this way a maternity roost community can leave its roost together and seek out another one. In order to be able to cling tightly to the mother, the hind feet and thumbs are already 80% of their ultimate length at birth; on the other hand, the forearm, which is not as important, is only 30 to 40% of its ultimate length.

Young bats develop very rapidly. Depending on the species, the eyes open at three to ten days of age. Newborns have sparse, short, barely pigmented hairs. The actual fur starts growing during the first week of life. When they are only a few days old the young can already run fast and climb very well, and thus are by no means as helpless as they might appear at first glance. The young of the Common Bats have well-developed milk teeth, which are lost in the first few weeks. The permanent teeth begin to erupt starting on the tenth day. The young are capable of flight at three to four weeks of age (Nathusius's Bat and Common Pipistrelle). Larger species, such as the Mouse-eared Bat and Noctule, are able to fly at an age of five weeks. The young are weaned from the mother's milk after four to six weeks and must then catch insects independently. After weaning their young, the mothers seek out breeding roosts. The young do not leave the maternity roost until some time later. In bad weather the young are in particular danger at this time, because they have few fat reserves and are still inexperienced hunters.

The mortality rate of the young in this critical phase can be as

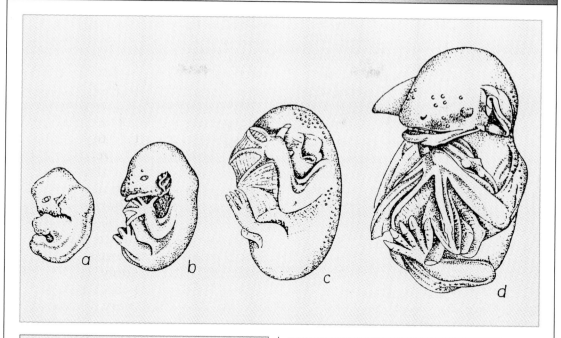

Fetuses of a Mouse-eared Bat of different ages.
 a: 9 millimeter skull-coccyx length
 b: 13 millimeter skull-coccyx length
 c: 17 millimeter skull-coccyx length
 d: skull-coccyx length unknown
Stage c shows the beginning of the development of the front extremities into an organ of flight. The wings are still quite underdeveloped at birth; on the other hand, the hind legs are well developed as grasping organs (d).

Changes in the proportions between the body and wing sizes in young and sub-adult bats (Noctule).
The wings, which are still underdeveloped at birth, grow within three to four weeks into functional organs of flight.
 a: 1 day old
 b: 28 days old
 c: fully grown (approximately 60 days old)(from Mohr, 1932).

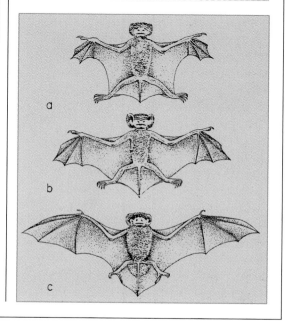

high as 50% in some species. It is estimated that only 30 to 40% of the young reach their second year of life. However, the mortality rate is considerably lower in the following years.

At birth, the sex ratio of bats is in balance at 1:1. The mortality rate of males is somewhat higher than that of females, which does not have a negative effect on the survival of the species because of their reproductive biology (harem formation).

THE SOCIAL BEHAVIOR OF THE BAT

Although the social behavior of many bat species has not yet been studied in detail, the basic behavioral patterns appear to be the same or similar in individual genera and species of a family. Many behavioral patterns, such as flying, echolocation, prey capture, and hibernation, have already been described.

In the following section, a number of unusual behavior patterns will be described briefly. Basic behavioral patterns, such as grooming, stretching movements, the manner of defecation and urination, wing movements, and so forth, can already be observed in youngsters in the course of the first and second weeks of life. Therefore, they are thought to be instinctive behaviors.

As pronounced social mammals, bats live together in groups for the majority of the year. Communities of different species are very common in the winter roost, but also occur in maternity and other roosts. A hierarchy within a group of conspecifics has not yet been described in European bats. Such a hierarchy is most likely to occur among the males during the mating season. Common Bats hang densely packed or even on top of each other and often disturb one another in the roost. When this happens there is often a confrontation wherein they open their mouths wide, revealing their sharp teeth, scold shrilly, and strike out at the troublemaker. The situation hardly ever escalates into serious fights or even injuries. Order is usually soon restored. The facial expression is very similar in both "attacker" and "defender," although in the defensive animal the ears are laid back slightly. If, on the other hand, a conspecific is attacked, the ears are erect. Bats behave similarly when they feel threatened by a human hand and cannot escape. Larger species then bite the hand very hard and painfully. Smaller species, such as the Common Pipistrelle, cannot break the human skin.

In Nathusius's Pipistrelle, the Common Pipistrelle, and probably other members of the genera *Pipistrellus* and *Nyctalus* (*N. noctula*), a special form of defensive behavior can be observed. If the external threat is not very intense, the bat presses itself flat on the ground in a fright posture. However, if it is touched or even placed on its back it curls up and allows itself to be manipulated as if it were lifeless. This state, which is called akinesis, is particularly easy to trigger in somewhat lethargic animals and represents a "playing dead" reflex.

As in many social animals, a

Facing page: Top: With the akinesis of this Nathusius's Pipistrelle (*Pipistrellus nathusii*) it is a question of a "playing dead" reflex. Even when the bat is picked up it remains motionless on its back. The thick fur on the tail membrane is clearly visible. Bottom: Bats groom themselves primarily with the claws of the hind feet. This Barbastelle (*Barbastella barbastellus*) lifts the foot over the wing to the head. The eyes are closed during this time.

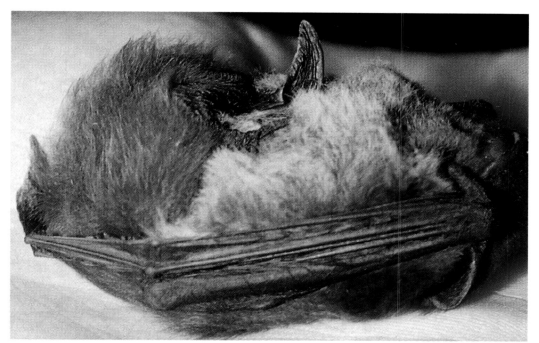

"join in" behavior can also be observed in bats. If a bat takes off, grooms itself, or goes to the food bowl in captivity, this has the effect of stimulating its conspecifics in the same direction. Mutual grooming does not occur or occurs only rarely among adults of the European Common Bats. Mothers also lick their young for only a few days, after which the youngsters groom themselves. It has been observed in the Mediterranean Horseshoe Bat that even adult individuals lick one another on the face and head. When doing this they hang belly to belly, and one individual embraces the other with its wing membranes.

Body care takes up a great deal of time in bats of all genera. They groom themselves regularly before they leave the roost and after hunting as well. The bats thoroughly lick their wing membranes, which they pull over their face like a mask while grooming. They are oiled and kept sleek by the secretions of the facial glands. The fur is combed with the claws of the hind feet. During grooming, the claws are repeatedly licked and cleaned at short intervals. During body care, the bats hang by only one leg and pass the other either under or over the wing to the fur. The wings are repeatedly spread wide. A shaking of the entire body can also be observed. The areas of fur on the chest and belly that can be reached with the tongue are also licked. Like other mammals and birds, bats can also yawn.

The fact that the innate behavioral elements are firmly fixed is demonstrated when a bat is fed on the ground. Because almost all species catch their prey in flight, they must first learn to take food on the ground. In so doing, they exhibit the same behavior as in flight. For example, they quickly seize a mealworm and try to push it into the pocket formed by the tail membrane. Sometimes they tumble over and lose the prey. In contrast to the Common Bat and the European Free-tailed bat, horseshoe bats, because of their build, can only eat and drink in flight or while hanging. When excited (for example, while catching a prey animal) and during and after food intake, feces and urine are excreted. This can also occur in flight in all species.

Characteristic of Common Pipistrelle is the practice of sticking their feces on walls, window panes, and other surfaces, which they touch briefly in flight. This "feces sticking" is so typical of the species, it makes it easy to detect their presence. When common bats defecate and urinate while hanging, they first rotate with the head upward and raise the tail membrane to protect it from being soiled. Because the hanging horseshoe bat has its tail with the tail membrane folded onto the back and not onto the belly as in the common bat, there is no danger of soiling.

Facing page: Grooming behavior is an instinctive behavior of bats. Here a six-day-old Common Pipistrelle (*Pipistrellus pipistrellus*) licks its wing membrane.

When urinating the common bats arch the tail membrane upward. With this male Common Pipistrelle (*Pipistrellus pipistrellus*) a drop of urine is visible on the penis.

In closing, we mention a behavior that can cost the life of many bats that enter buildings in their search for roosts. Common Bats and Horseshoe Bats fly up to a selected place several times before landing. These exploratory flights can also be observed in roosts that are very familiar to the bats. Horseshoe bats always hang on the perch, but Common bats also exhibit a kind of "belly landing." The bats fly over the hollow space they have selected (vases, glasses, lampshades that are open on top, open pipes, double-hung windows, etc.) several times and then let themselves drop in. This type of landing can frequently be observed in the Common Pipistrelle. It is also encountered in the Common Long-eared Bat, Daubenton's Bat, and other species. If the walls of the container are smooth and its diameter is small, the bats can neither climb nor fly out and will die.

Roer reports on such a bat trap, a ventilation pipe, that contained 1180 dead Common Pipistrelles. This example shows how the bats, which are so wonderfully adapted to their natural environment, can fail when confronted with the

Facing page: Top: Daubenton's Bats (*Myotis daubentoni*) often seek out narrow cracks in the winter roost. Bottom: Many bat species exhibit great adaptability in the choice of their hanging places in the winter roost. Here a Daubenton's Bat (*Myotis daubentoni*) has hidden itself in the root-filled soil of an underground tunnel.

"creations" of human civilization.

HIBERNATION—LIFE ON THE PILOT LIGHT

In the temperate zones of the earth, the winter months present special problems for some members of the animal kingdom. To survive the cold and lack of food requires special adaptations. Cold-blooded (poikilothermic) animals, such as amphibians and reptiles, are largely dependent on the ambient temperature. They cannot actively raise their body temperature substantially above the ambient temperature. They survive the winter in frost free hiding places in a cold-induced torpor, waiting for the warm rays of the spring sun to "bring them back to life." Most mammals, on the other hand, maintain a constant body temperature independently of the ambient temperature. They are warm-blooded (homoiothermic) animals. In the winter they protect themselves from the cold with thick fur or, like the mole, the shrew, and the mouse, take shelter in protective dens underground. In any case, they must have sufficient food available, because they would freeze to death without a constant supply of energy.

Birds, too, are homoiotherms. Many species escape from the winter by migrating to warmer regions where they find sufficient food. Some of the bats that live in the temperate zones also migrate in the fall to regions with more favorable climates, but even there the winter can affect them and completely wipe out the insect life that is the basis of their diet. All European bats, in the same way as certain other mammals (for example the hedgehog, marmot, and dormouse), have evolved a special survival strategy. In the fall they accumulate large amounts of brown fatty tissue and, with the aid of this energy reserve, they can sleep through the cold and foodless time. We find deposits of brown fatty tissue in bats especially between the shoulder blades, but also on the neck and sides. Because of these deposits, their body weight in the fall is about 20 to 30% higher than in the spring. In contrast to the cold-blooded animals, however, they are not completely at the mercy of the cold. They control their body temperature as if with a "thermostat," and can raise it, without an external supply of energy, to the normal level again. Animals with this ability are called heterotherms. Even in the warm season, bats save energy by largely matching their body temperature to the ambient temperature during the daily torpor. Eisentraut describes this state as "daytime sleep lethargy." In this phase the bats cannot fly immediately in cool weather (body temperature below 20°C). They must therefore raise their body temperature to normal before the start of the evening hunting flight. Studies of Mouse-eared Bats have shown that the young behave as cold-blooded animals in the first two days of life; thus they cannot actively regulate their body temperature.

In daytime lethargy and hibernation, long-eared bats fold back their large external ears and clasp them under the forearms. Then only the tragus of each ear points forward. On the left is a Gray Long-eared Bat (*Plecotus austriacus*) and on the right is a Common Long-eared Bat (*Plecotus auritus*).

Therefore, they are warmed by the mothers. Only when the fur growth of the young is completed at an age of about a month do they behave as full heterotherms.

As the cold nights increase in the fall, the organism of the bat slowly shifts to hibernation. Starting in October or November the bats seek out winter roosts. The precise timing is dependent on the outside temperature, the respective species, and apparently also on the internal clock.

Each species has a particular preferred temperature for the place they hang in hibernation. All require the humidity to be as high as possible in the roost to guard against dessication. The air is often saturated with water vapor, causing the bats to be covered completely with drops of dew. According to the choice of hanging places in the winter roost, we can distinguish between three groups of bats.

Species of the first group are always free-hanging by the hind feet on the ceiling (horseshoe bats). The second group can also be found free-hanging on the ceiling or the wall, but often seek out niches, shafts, or crevices with a more favorable

Left: Because of the high humidity in the winter roost, bats, such as this Daubenton's Bat (*Myotis daubentoni*), are often covered with large drops of dew.

Facing page: Greater Mouse-eared Bat (*Myotis myotis*) in the typical hibernation posture of a free-hanging common bat. The wing membranes shroud the body only partially; the tail is folded over the belly.

Below: The fully extended legs show that this Lesser Horseshoe Bat (*Rhinolophus hipposideros*) is hibernating undisturbed in a limestone cave. When hibernating, this species always shrouds itself completely in the wing membranes.

microclimate (for example, the Mouse-eared Bat and Daubenton's Bat). The species belonging to the third group (for example, the Common Pipistrelle) as a rule prefer narrow cracks, in which they can have body contact with the walls on all sides. There is great variability in the choice of hanging places in the second and third groups; where the bats ultimately hang probably depends on the opportunities in the roost and especially on the microclimate.

Bats additionally reduce heat loss during their daytime sleep lethargy by using particular body positions. Horseshoe bats wrap themselves in their wing membranes like a cloak. Long-eared bats fold back their long external ears, which are well supplied with blood, and tuck them under their wings. All other common bats press their wings and tail membranes tightly to the body, thus reducing their body surface.

Complicated hormonal regulatory mechanisms play a role in the far-reaching changeover of all bodily functions, particularly the metabolism, to the "pilot light" of hibernation. They bring about a drastic lowering of the heart rate and breathing frequency as well as a reduction of the body temperature. According to studies by the zoologist Kulzer, the heart rate of the Mouse-eared Bat reaches up to 880 beats per minute when excited, in the resting phase it is 250 to 450, and in deep hibernation it is only 18 to 80 beats per minute. The breathing frequency is 4 to 6 breaths per second in the waking phase. It is drastically reduced in hibernation, and there may be pauses in breathing of 60 to 90 minutes, making the bats appear dead. Active Mouse-eared Bats have a body temperature of about 40°C. In deep hibernation, on the other hand, their body temperature is between 0°C and 10°C. Hibernating bats as a rule set their body temperature 1-2°C higher than the ambient temperature. This great cutback in all of life's processes achieves a substantial energy savings.

In medicine, as well, the energy saved from lowering the body temperature has been used, without even distantly achieving the perfection of bats. Because humans are homoiotherms, external cooling prompts immediate counter regulation. The body temperature is kept constant by elevating the metabolic rate. With certain medications it is possible to switch these counter regulatory mechanisms off by blocking the autonomous nervous system, and ultimately achieving body temperatures of 20°C and lower by cooling the body with, for example, ice. In this state of hypothermia, oxygen consumption in humans falls to 25 to 30% of its starting value. As a result of this substantial cutback in the metabolic rate, it is easier to treat illnesses with high fevers and to perform complicated operations.

When a bat awakens from

hibernation spontaneously or because of an external stimulus, the respiratory and heart rates increase rapidly. This increases the oxygen supply to the tissues and heat is produced slowly. About 30 to 60 minutes after awakening, the bats have reached their normal body temperature. The lower the body temperature in hibernation and the lower the animal's energy reserves, the longer it takes the animal to awaken from its torpor. In the initial phase of awakening, the production of heat occurs predominantly by metabolizing the brown fatty tissue. In the

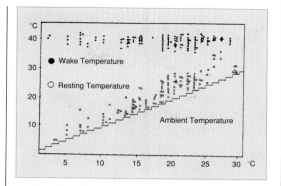

Resting and waking temperature in the Mouse-eared Bat The measurements of numerous animals show that the resting temperature is only a few degrees above the ambient temperature (from Kulzer, 1981).

When awakening from the state of torpor, the heart frequency (Hf) increases ahead of the rise in body temperature (Kt). It initially exceeds substantially the normal resting rate in the waking state and settles down to this level in a short time (Mouse-eared Bat, from Kulzer, 1967). UT = ambient temperature.

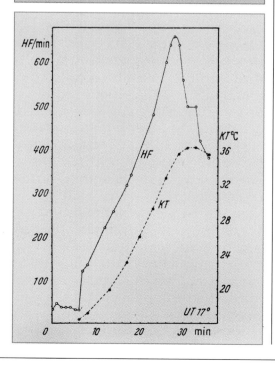

second phase, heat is produced predominantly by a clearly visible muscular shivering.

A bat is helpless in deep hibernation. It is able to react reflexively, but only very slowly. When touched, bats pull-up their legs, open the mouth wide, and emit persistent, high-pitched, defensive calls. They bite reflexively in this situation and keep their jaws closed for a long time. If you slowly free the animal from its substrate, it spreads its wings and falls, as if hanging from a parachute, with circling movements to the ground. They can turn over from the supine position only with great effort. On the other hand, they are considerably better at clinging to a substrate. Only as their body temperature slowly rises are they gradually able to take control over all their bodily functions again, before finally being able to fly away.

During hibernation slight touching, extended lighting with a

In contrast to the common bats, in the resting phase the horseshoe bats fold their short tail over the back. This Lesser Horseshoe Bat (*Rhinolophus hipposideros*) displays a protective reflex when disturbed in the winter roost: with a pull-up it slowly raises up on its hanging place.

flashlight, or frequent photography with a flash, act as waking stimuli. If the temperature in the winter roost falls below the preferred temperature of the respective bat species, the animal is also awakened. The animals awaken and seek out a place to hang that has a more favorable microclimate. They occasionally even change their roost in the winter. Extended waking phases in the winter, however, are dangerous for the bat, because its energy reserves will be used up too quickly. There is then the danger that they will no longer have enough energy for the waking process in the spring, or that they will not be strong enough for the necessary food acquisition. The loss of warmth, and hence energy, in hibernation can be reduced by forming groups (clusters). The animals then hang either close together or even on top of one another like roofing tiles. As a result of this social thermoregulation, the bats on the inside of such groups lose less heat than solitary, free-hanging individuals. Apart from the horseshoe bats, which always hang freely at a distance from their neighbors during

Right: Large cluster of Greater Mouse-eared Bats (*Myotis myotis*) free-hanging from the ceiling of a concrete tunnel.

hibernation, cluster formation is possible in all European bat species. There are clusters that are formed by only a single species, but up to five species

Mixed cluster of Daubenton's Bat (*Myotis daubentoni*), a Greater Mouse-eared Bat (*Myotis myotis*) above right, and Barbastelles (*Barbastella barbastellus*) to the left.

Bats are largely helpless in hibernation. This Greater Mouse-eared Bat (*Myotis myotis*), which has been placed on its back, has spread its wings in reflex and performs slow grasping movements with the feet.

Bats do not sleep continuously in hibernation, but awaken spontaneously several times. The length of the individual sleeping phases is dependent on its internal clock, the temperature in the winter roost, the period (start, middle, end) of hibernation, and the respective species. Sleeping phases can last from a few days up to one or two months. During the brief waking phases, the animals fly around in the winter roost, defecate and urinate, and possibly also drink water and

have also been observed in such gatherings, although one species usually dominates. Cluster formation of this kind is typical, for example, with the Mouse-eared Bats, the Noctule, and the Common Pipistrelle.

catch prey animals. In some species, such as Daubenton's Bat and the Barbastelle, mating can also occur in the waking phases. Hibernation comes to an end in the spring (March or April). The precise time depends on the internal regulatory mechanism and the outside temperature.

ON MIGRATION TO FROST-FREE WINTER ROOSTS

Bats undertake migrations of various lengths between their summer and winter roosts. We can therefore distinguish between migrants, partial migrants, and permanent residents. Typical representatives of the migrant species are the Noctule and Nathusius's Pipistrelle, which, as a rule, leave their summer biotope and often cover distances of more than 1000 kilometers to reach their winter roosts located further to the south. Partial migrants, such as Daubenton's Bat and Mouse-eared Bats, cover distances of more than 100 kilometers on the way to their winter roost, whereas resident species (for example, horseshoe bats, long-eared Bats, and Common Pipistrelles) usually do not exceed distances of 20 to 50 kilometers. Because migratory species seek out regions with a milder climate, a distinct direction in their migration can be determined. The partial migrants and permanent residents, on the other hand, arrive at their winter roosts from all directions. It is observed repeatedly that individuals of the

Migration routes and recovery sites of Noctules (*Nyctalus noctula*) banded in Germany. Solid arrow: animals banded in the summer roost. Dashed arrow: animals banded in the winter roost (from Heise and Schmidt, 1979).

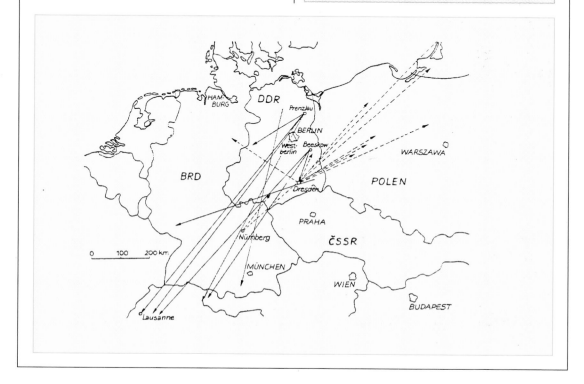

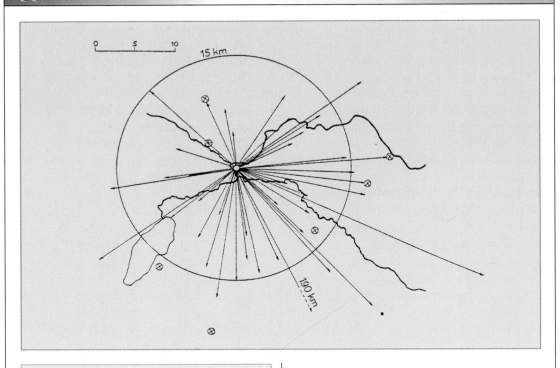

Migration routes of Common Pipistrelles (*Pipistrellus pipistrellus*) that were banded in the winter roost in the Demmin church (Mecklenburg, East Germany). Localities with maternity roosts are marked with a cross (from Grimmberger and Bork, 1979).

latter two groups also cover distances that are far above average for their conspecifics. The reasons for this are, for the most part, still unknown. Moreover, in the total range of a species, the migratory behavior of individual local populations can vary. All populations living at the northern boundary of their range cover longer distances than the more resident populations of the same species living farther to the south.

The record holders among migratory bats are Nathusius's Pipistrelles and Noctules, which travel distances of up to 1600 kilometers.

BANDING

The migrations of bats can only be explained by marking the animals. Eisentraut, in 1932 in Germany, was the first to band bats in Europe. The rings (which actually are clamps) manufactured for this purpose are applied to the forearm of the bat in such a way that it is still freely movable. The rings are supplied in various sizes by the central banding offices of the federal states, and carry the registration number and the abbreviated address of the central office. Depending on the type of ring, the weight varies from 0.10 to 0.19 grams, so they do not interfere with the ability of the bat to fly. The bats, however, at first consider the ring to be a foreign body. They bite it, and, if it is made of an unsuitably soft material, it becomes squeezed tightly together, which can cause injuries to the wing membrane. Because of the efficient regenerative ability of the wing membrane, such an injury usually heals well. Banded bats with maximum ages of more than 20 years and successful migrations of more than 1000 kilometers lead us to the conclusion that banding is not particularly dangerous to the severely endangered species may be banded. Since the introduction of bat marking in Europe in many species, well over ten thousand individuals have been banded. This impressive number has contributed to answering many questions. From the wealth of data, the following are singled out:

—Clarification of the migrations between the

Bat ring (arm clip). The bat ring is attached to the forearm and carries the most important data of the banding center.

animals. Nevertheless, we must seriously consider whether it is still necessary today to band every individual in every population. Banding should only be done if it helps us to answer specific scientific questions. The results should also serve, directly or indirectly, for the protection of the animals. It is basically forbidden to band in the maternity roosts, and no

individual types of roosts, specifically between the summer and winter roosts.

—Homing ability.

—Site faithfulness.

—Changing of hanging sites in the winter roost, duration of the sleeping phases (marking with colored rings).

—Age structure of the population, maximum age, reproductive age.

For many central European species, therefore, no substantial increase in knowledge can still be expected through banding.

What should you do when you find a banded bat?

—If a live, banded bat is found, it should be released immediately after recording the information on the ring.

—The ring number should be reported to the central banding office or the nearest conservation office along with information about the circumstances under which the bat was found, and, if possible, the species and sex of the animal.

—With a dead bat, the ring should be removed and sent to the central office along with the report.

—If possible, the dead animal (preserved in alcohol) should be donated to a zoological museum as a scientific specimen.

If capturing the bats is necessary (for example, destruction of the roost, banding), with permission of the responsible conservation authorities, the following roughly outlined methods of capture can be used. In the winter roost, you can (cautiously) pick the bats relatively easily from the ceiling or pull them out of crevices if they are not too deep. If the animals hang from a fairly high ceiling, an assistant should spread a "safety blanket" below the animal.

It is considerably harder to catch bats in the summer. It is possible to catch bats in front of cave entrances, over smaller ponds and river courses, and on paths or firebreaks, with a not too taut Japanese net spread across the flight path. To choose the right location, it is advisable to observe the flight paths of the bats on the previous evening. The net must never be left unsupervised, because otherwise the bats could become badly tangled and injure themselves. Larger species can also bite holes in the net very quickly.

As bats fly from their roost (tree holes, attics, and other places), we can catch them with a plastic sack or gauze spread in front of the opening. The trap devised by Constantine, consisting of a rectangular frame covered with nylon threads (threads 5 cm apart, thread thickness 0.1 mm), has also proved to be very effective. The lower third of the frame is inserted in a sufficiently deep plastic sack fastened to another frame. The upper part, resembling a harp, is installed in front of the entrance hole. The bats fly into the threads and slide down them and into the plastic sack, from which they cannot escape. Here, too, the trap must be constantly supervised. If the places where the bats hang in the roost are accessible, then it is also possible to take them from there (but not in a maternity roost with dependent young!). When catching bats in their roost, you should choose cool days, otherwise the animals will be very active and large species may also bite.

Remember, when picking up a bat, you should always wear gloves or use a thick cloth.

Brandt's Bat (*Myotis brandti*) in a Japanese net.

PROTECT OUR BATS!

European bats have only a few natural enemies. Among the birds, the Tawny Owl (*Strix aluco*) and the Barn Owl (*Tyto alba*) prey on the bats as they emerge from their roost or in a maternity roost. Species that fly in late afternoon or in early evening can also be preyed on by falcons. Among the small predators, the Beech Marten (*Martes foina*) has been documented as preying on bats. Domestic cats usually only kill bats and then leave them—as they do with shrews. In countries outside of Europe it has been observed that some snakes (for example species of the genera *Coluber* and *Elaphe*) have specialized in catching bats at their roosts or as they fly by.

The ectoparasites (fleas, bugs, bat flies, ticks, and mites) that live in the fur and on the wing membranes of bats, are certainly more or less burdensome to their host, but probably are dangerous only in exceptional cases, for example to a sick or weakened animal. As carriers of diseases that can also infect humans, the European bats—in contrast to certain rodents—as a rule do not play a role. Because they can, however, in exceptional cases, as with all mammals, be infected by rabies, bats—particularly those that cannot fly or that behave abnormally—should only be handled with gloves. This is particularly true of large, powerful species!

It is not the few natural enemies that have threatened the populations of many bat species for the last several decades, but rather it is human civilization. With its modern industrial complex that so severely endangers the existence of bats, humans civilization has directly or indirectly contributed to the extermination of some bat species (at least regionally). According to studies by Roer, the Greater and the Lesser Horseshoe Bats disappeared from their range in western Central Europe in the 1960s. In some regions reproduction can scarcely still be documented, so it is expected that these species will eventually die out there. With the Mouse-eared Bat, a population decline has been recorded as well, particularly in the northern part of its range. There are populations that currently are at only 20% of their former levels; in various places the maternity roosts have already been lost.

It is to be hoped that at least regionally larger populations will find optimal living conditions and that individuals will live long enough to ensure the preservation of the species.

THE FOLLOWING SPECIFIC FACTORS ARE RESPONSIBLE FOR THE DRASTIC DECLINE OF BAT POPULATIONS:

—Decrease in or destruction of the food supply through the use of pesticides and direct poisoning of the bats through poisoned, but still living, food insects.

—Destruction of natural landscapes and living spaces.

—Destruction of bat roosts by the demolition or modernization of old buildings, the hermetic sealing of attics, the destruction or complete sealing of underground hollow spaces and old cellars, and the cutting down of hollow trees.

—Use of wood preservatives that are highly toxic to warm-blooded animals in bat roosts (for example, in roof trusses).

—Serious disturbance and killing of bats by humans.

—Accidental deaths in containers or pipes that are open above but act as bat traps.

—Unfavorable climatic factors (increased mortality of bats that are awakened from hibernation and of young in protracted periods of cold and wet spring or summer weather).

Most bat species in a particular region are exposed more or less equally to these factors. Nevertheless, it is noteworthy that not all are affected equally. Populations of Natterer's Bat, the Common Long-eared Bat, and the Common Pipistrelle have scarcely exhibited a decline. In recent years a population increase has even been observed in western and central Europe for Daubenton's Bat. On the other hand, the Hardy Barbastelle has declined just as drastically in many regions as the aforementioned warmth-loving Mouse-eared Bats and Horseshoe Bats. There has been no definitive explanation for the different reactions of the various species so far. It is conceivable that the prey insects of the individual species are being affected to a different degree by the use of insecticides and the destruction of the natural countryside.

All bat species are protected by law; in some countries they are present on the endangered species lists of animals that are threatened with extinction. Yet legal protection alone is not slowing the decline! The protection of bats, as is true of the protection of all animal species, must begin with the protection of the biotope. We must preserve their natural habitat and their roosts.

Thus, it is essential to protect their winter roosts from unsupervised visitors. The installation of grilles with horizontal bars at tunnel entrances or an entrance hole (50 to 75 centimeters wide, 15 centimeters high) in walled-in entrances to subterranean spaces make it possible for the bats to have continued unimpeded access to their roosts. In winter roosts with walls that are too smooth, hanging places can be

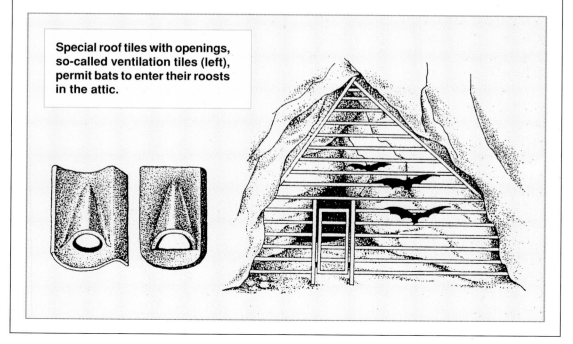

Special roof tiles with openings, so-called ventilation tiles (left), permit bats to enter their roosts in the attic.

provided by attaching hollow building blocks or boards (at a distance of 2 to 4 centimeters from the wall) to the ceiling or the walls. To improve the microclimate, it is advisable to close off or reduce the size of too large openings in the ceiling or entrances to the winter roost.

To preserve the summer roosts in attics, the entrance holes must be retained during renovation of the roof. Suitable for this purpose are ventilation tiles installed in the roof, from which the screen insert is removed, or clay pipes left in gabled roofs. Horseshoe bats require an open entrance, the opening of which should be about 20 x 30 centimeters. It is important, whenever possible, to deny feral pigeons access to the attic spaces in which the bats live.

Sometimes the homeowner is disturbed by the bat droppings under the roost. Here a plastic tarp spread or stretched out under the roost can provide relief. Also keep in mind that bat guano is a valuable fertilizer for flowers and gardens.

In known maternity roosts, roof work and measures to preserve wood should be carried out only in the period from October to February. If possible, no changes should be made to the hanging places.

Because wood preservatives with chlorinated hydrocarbons (lindane, PCP, hylotox) have a toxic effect on bats for months and even years, it is advisable not to use these compounds. Instead, use compounds that are harmless to mammals, such as preparations based on permethrine.

The control of wood pests using the hot-air method should also be considered. If the nontoxic methods are not feasible, then we should try to cover the hanging places with old, unimpregnated boards before using pesticides in the attic.

Places where bats live outside on houses, behind shutters or wooden blinds, should be retained if possible. As a substitute or to supply additional roosts, bat boards (positioned 2 to 3 centimeters from the house wall) can be installed on a sunny wall.

For the forest bat the preservation of trees with holes is of paramount importance. In recent years, hanging bat boxes on trees as substitute holes have also proved effective. This practice has not only led to the first confirmed reports of bats in particular regions, but in individual cases has also led to true new colonization by bats.

Is hard to say which of the various types of boxes of wood or reinforced concrete composite material are best suited, because bats exhibit great adaptability. The FS 1 box developed in eastern Germany by Stratmann has proved to be particularly effective. This narrow box uses inexpensive materials and is easy to build. Maternity roosts of Nathusius's Pipistrelle, the Common Pipistrelle, Leisler's Bat, the Noctule, the Common Long-eared Bat, and Brandt's Bat have

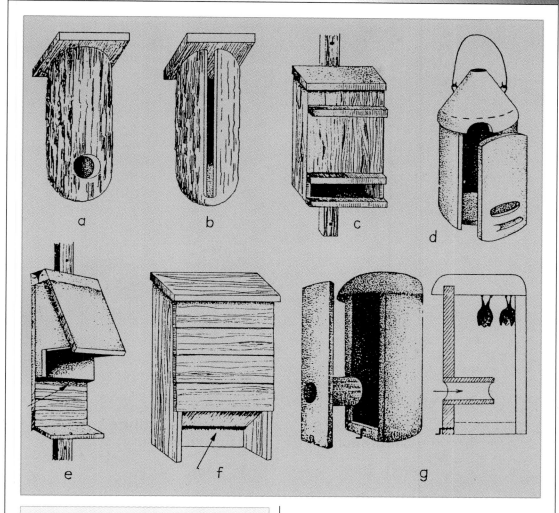

Various types of bat boxes that are suitable as artificial summer roosts for forest bats.

a: Hollowed-out section of trunk (corresponds to the upper part of a
   woodpecker hole).
b: Hollowed-out section of trunk with a
   slit-shaped entrance.
c: Reinforced concrete composite
   (Wonderboard®) box (model from
   Issel).
d: Reinforced concrete composite box
   (model from Schwegler, 2 FN).
e: Wooden box (Dutch model).
f: Wooden box (model from Stratmann,
   FS 10.
g: Reinforced concrete composite cy-
   lindrical box (model from Nagel).

been found in it. Individual instances of the use by Daubenton's Bat, Geoffroy's Bat, the Serotine, and Mouse-eared Bats have also been documented. Refinements in the box have been achieved by covering it with unsanded roofing felt, as proposed by Heise, and covering it with sheet metal with dark paint, as devised by Issel. These refinements produce a more favorable microclimate because of the increased absorption of heat. Moreover, the boxes are largely protected from damage by the

Greater Spotted Woodpecker. When building the box it must be kept in mind that the boards must be about 15 mm thick. They must not be impregnated with wood preservative and must remain rough on the inside.

At a height of three to five meters, they are fastened, if possible on the sunny side of the trunk, with light metal pins, wood screws, or left free-hanging on a wire hanger. Because forest bats change their roost frequently, five to ten boxes should be installed 20 to 50 meters apart on the

Occurring in this mixed birch-conifer forest are the Noctule (*Nyctalus noctula*), Nathusius's Pipistrelle (*Pipistrellus nathusii*), and the Common Pipistrelle (*Pipistrellus pipistrellus*).

edges of paths, firebreaks, or clearings. Even if the boxes are not occupied by bats, they are often used by other beneficial animals. For example, Tree Creepers have been known to nest in FS 1 boxes. Blue Tits have nested in other types of boxes, but hornets or wasps sometimes colonize them as well.

When bats come into human hands by accident or because of the destruction of their roost, they must be released immediately. If the animals give a weakened or sick impression, they must be brought to a veterinarian, a zoo, or a similar institution while observing the aforementioned precautionary measures. In the winter they can survive in an artificial roost with

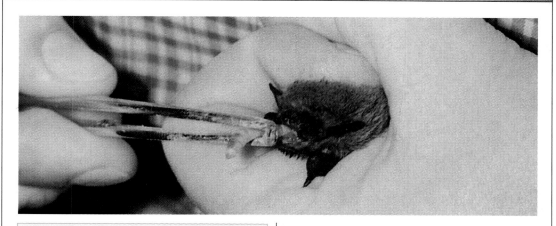

Most bats cannot eat at first under human care, but rather, as in this Common Pipistrelle (*Pipistrellus pipistrellus*), must be fed with forceps.

a constant air temperature of about +3 to 6°C and high humidity. Water must be offered for the times when the bats wake up temporarily.

In the summer, weakened or injured bats must first be given water to drink from a pipette. Mealworms (larvae of the meal beetle, *Tenebrio molitor*) are suitable for feeding. Initially, freshly molted or chopped mealworms, which have been killed first by squeezing their head with forceps, are offered. When feeding the bat or giving it water, it must be held in the hand so that the wings are pressed against the body and only the head sticks out. With larger species you should always wear leather gloves to protect against bites. With a certain amount of patience, you must stick a half a mealworm repeatedly in the bat's mouth. At first it will eat only the soft contents of the mealworm and reject the chitinous exoskeleton.

The bats usually learn after two to three days to eat and drink independently from a shallow bowl on the floor. On the other hand, horseshoe bats can eat and drink only while hanging.

Because a diet of mealworms is very one-sided, once or twice a week a multivitamin preparation and a mixture of minerals including calcium, similar to that used in animal breeding, should be administered. The supplemental feeding of flies, mosquitoes, and moths is very beneficial. The daily food requirement of up to a third of the body weight absolutely must be met! The bats must be able to hang in their roosts. The optimal accommodation for bats is an aviary; a fairly large terrarium should be used only in an emergency. In any case, the bats should be able to exercise daily. In the first days great patience is usually necessary. Once they have completely recovered, exhibit normal body weight, and display normal flight behavior with stamina, they should be released in good weather near the place where they were originally found.

Lesser Horseshoe Bat
in the winter roost.

WHICH BAT IS THAT?

EUROPEAN BATS

In the following section all European bat species are treated according to a fixed format. The discussion of subspecies or extremely rarely documented accidental species is avoided. The text is arranged as follows:

Distinguishing characters: Characters that are typical of all species of a genus are introduced in the brief description of the genus. The description of the external characters (for example, coloration and ear shape) is supplemented by the color photographs and the detailed photographs and figures that accompany the diagnostic key. When judging the ear shape the great flexibility of the ear should be taken into account. The photographs of the ears show the ear largely relaxed in the alert bat. Uncertain characters, such as the degree to which the ear projects beyond the tip of the snout when it is folded forward or the ratio of the thumb length to the wrist width, are not mentioned. Characters of the teeth must be evaluated with a magnifying glass. They exhibit a certain variability and are harder to determine in old animals with worn teeth.

Measurements: The measurements are derived from the main area of occurrence of the species. The data are presented in millimeters (mm) and grams (g). Extreme measurements are given in parentheses. The measurements of the individuals of a species are often lower in the southern part of the range than in the northern part (Bergmann's rule). Females are, on average, larger than males (difference in the length of the forearm approximately 1 to 1.5 mm). The body weight is up to 30% higher in the fall, before the start of hibernation or with highly pregnant females, than at the end of hibernation or with a female that is not pregnant.

Anomalies in coloration: As far as is known, cases of anomalies in pigmentation, such as partial or complete albinism, are presented.

Similar species: The sequence of species corresponds approximately to the likelihood of confusion. Individual characters must be referred to in the description of the respective species or in the diagnostic key.

Range: The data serve only for orientation. For some species, the knowledge of the range is still very sketchy.

Biotope: In the descriptions of the biotope, the great adaptability of the individual bat species to different geographic and climatic conditions must be noted. The same is true of the temperatures in the winter roosts, the type of hanging places, and the maternity roosts.

Migration: An exact classification of the migratory behavior is not possible with the species that have been less studied. It must also be considered that even individuals of normally sedentary species in exceptional cases can undertake long migrations. The data on the longest known migration do not always refer to a single season (summer/winter); sometimes a number of years lie between banding and recovery.

Reproduction: Dates, such as the start and end of hibernation, formation of the maternity roosts, and the birth of the young, vary depending on the climate. Note that with the data on the birth dates, the young are born considerably later after a cold spring. Because gestation periods vary greatly because of their dependence on climate, they are not given.

Maximum age: The maximum age attained so far is given for the respective species. The actual maximum age is probably somewhat higher. In any case, the average life expectancy is considerably lower than the maximum age.

Hunting and diet: Prey lists, which already exist for a number of species, have intentionally been avoided. The time of emergence from the roost, the duration of the hunting flight, the flight altitude, and the type of hunting territory can vary greatly. They are substantially dependent on the season, the weather, and the kind of prey animal. Species identification by the flight behavior is not possible with most species, but the supplemental use of a bat detector increases the likelihood of correct identification.

Calls: A description of the audible calls is considerably more difficult in bats than in birds; therefore, the data serve only for orientation. As the result of the development of transportable, electronic ultrasound detectors (so-called interference detectors, or tenth frequency detectors), it has become possible to hear the calls that flying bats emit for navigation, prey detection, and social communication. In this way it is possible to detect bats in darkness and to better study their behavior in flight.

Field studies of various ecological questions (for example, selection of hunting habitats) and the analysis of the acoustic repertoire in connection with the hunting behavior, however, require patient training.

Because the price of the detector models offered on the market is quite high, before their purchase you should consult with an expert to determine the pros and cons of the individual detectors. It absolutely is not true that you can determine immediately which species has just flown by with the press of a button.

Studies that have been published so far have shown that many species can be identified with some practice on the basis of such criteria as the call form, call intensity, call duration, frequency range, and the repeat rate of the calls.

The Horseshoe Bat species differ from the Common Bat species in that they transmit pure tones of relatively long duration (up to 60 ms), so-called constant frequency calls (CF calls). The Common Bats, on the other hand, primarily produce frequency modulated calls of short duration (FM calls of from about 12 kHz to

100 kHz, of 2 to 200 ms duration), which in some species can also have CF components in certain situations. Within the Common Bats it is particularly difficult to distinguish between the species of the genus *Myotis*. Apparently it is not possible to distinguish between *Myotis mystacinus* and *Myotis brandti* at all. The aforementioned criteria are subject to many influences, so that it is recommended at first to combine the acoustic observations with visual observations.

Particular significance is given to distinguishing between the search, approach, and capture phases in the hunting flight: In the search phase (uniformly controlled, usually direct flight), the calls are long (2 to 20 ms), as are the pauses between the calls (small species 85 ms, large species up to 300 ms). When a prey animal is located, the approach phase begins. The calls and the pauses between them become shorter. Finally, in the capture phase, very short calls are transmitted (in part only 0.5 ms in duration), and the pauses in between can be shortened to as little as 5 ms.

It is customary to represent the sound picture in the search phase, which can be in the form of an oscillogram (wave form [amplitude] and duration), sonogram (frequency range and duration of the call), frequency spectrum (data about the maximum volume in relation to the frequency [two dimensional], as well as the time [three dimensional]) or impulse repeat rate (number of calls per time interval). Additional criteria that can influence the sound picture include:

A change in the biotope in which the bat hunts.

A change in the flight altitude (*Nyctalus noctula* exhibits a different sound picture when hunting above the treetops than when hunting a few meters above the ground).

Hunting in groups or individually. These different factors can have the effect that often two or more call types are used regularly, which sometimes can cause confusion.

The distance over which the calls can be detected depends on the species and its types of calls, and on the sensitivity of the detector and the position of the microphone in relation to the approaching bat.

Protection: All bats are threatened by the destruction of their living spaces, particularly the summer and winter roosts, the extermination of their prey animals through the use of pesticides, or the poisoning of the animals themselves by wood preservatives that are toxic to warm-blooded animals. Bats are protected by law in almost all European countries. In most of these countries the law is very far reaching and very strict. The protection of the animals and their roosts takes precedence over research.

Banding and capture require official approval.

Winter and summer roosts may in part be visited only for the purposes of counting or of carrying out protective measures.

The animals may not be taken from their roosts and may not be photographed. Dead bats that are found should be handed over to scientific institutions (museums).

Additional figures: Refer to additional illustrations in the general section or in the diagnostic key.

HORSESHOE BATS, FAMILY RHINOLOPHIDAE GENUS: *RHINOLOPHUS* (LACEPÈDE, 1799)

Approximately 70 species currently exist, five of which live in Europe. The nose is surrounded by leaf-like skin processes, which are differentiated into the following parts:

> **Structure of the nose leafs in horseshoe bats (here *Rhinolophus euryale*) (illustration from Kusyakin and from Deblase).**
>
> **1: horseshoe, 2: saddle (sella), 2a: upper saddle process, 2b: lower saddle process, 3: lancet, 4: horizontal fold under the lancet, 5: nostril.**

Horseshoe: A horseshoe-shaped border of skin, open on top, surrounding the nostrils. The lower margin of the horseshoe is distinctly notched in the middle. Upper ends of the horseshoe reach almost to eye level.

Saddle (sella): Rises like an ax lying on its back above the nostrils and is surrounded by the horseshoe. The shape and length of the upper and lower saddle processes are important for species identification.

Vertical furrow: Runs between the saddle and lancet.

Lancet: Triangular, tapering to a point above. There are three overlapping depressions on either side between the base of the lancet and the upper end of the horseshoe.

The ears are broad at the base, tapering to a point above, and can be moved independently of each other. there is no tragus, and the antitragus is well developed. The eyes are small. The wings are broad and rounded. Their flight is

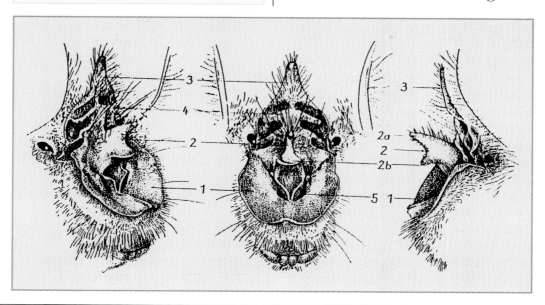

slow, partly fluttering like a butterfly, partly with short stretches of gliding, and extremely maneuverable. The tail is short and completely included in the wing membrane; at rest it is folded on the back. The spur reaches about a third of the length of the margin of the tail membrane. An epiblema is absent.

Females have two milk teats on the chest, and after the onset of sexual maturity develop two additional so-called false teats laterally at a distance of 2 millimeters from the genital opening. The young suck tightly to them in the first days of life. Females give birth to one youngster. Milk teeth are present only in the embryo; they are not present at birth.

Horseshoe bats are always free hanging in the roost, wrapping themselves totally or partially in the wing membranes while roosting. The youngster is also enclosed in the membranes in the first days of life.

Horseshoe bats barely go into daytime lethargy (torpor), quickly fly off, often hang by one leg, rotate around the longitudinal axis, and echolocate. They usually take off by dropping below the perch, but can also take off from the ground. They land after rotating the body axis by 180° with the head pointing downward.

Prey animals are frequently caught in the arm membrane, and can be stored for a short time in the cheek pouches. Larger food animals are eaten at fixed feeding roosts. In captivity, horseshoe bats seem better able to learn than European Common Bats.

Echolocation calls: Long frequency constant component, followed at the end by a short call of descending frequency (CF/FM call).

Species differentiation is in part possible by the frequency, because different species transmit in a narrow, species-specific frequency band. There are slight differences in the sending frequency within the same species depending on the geographic distribution. Calls are sent out through the nose.

$$\text{tooth formula } \frac{1\ 1\ 2\ 3}{2\ 1\ 3\ 3} = 32$$

RHINOLOPHUS HIPPOSIDEROS (BECHSTEIN, 1800)

Eng.: Lesser Horseshoe Bat
Ger.: *Kleine Hufeisennase*
Fr.: *Petit rhinolophe fer á cheval*
Head-body: 37-45 (47) mm
Tail: 23-33 mm
Forearm: (34) 37-42.5 mm
Ear (13) 15-19 mm
Wingspan: 192-254 mm
Condylobasal length: 13.4-14.5 mm
Weight: (4) 5.6-9 (10) g
Distinguishing characters: Smallest European horseshoe bat. It has a delicate build. The upper saddle process is short and rounded, the lower is distinctly longer, tapering to a point in profile. The fur is light gray at the base. The back is brownish smoky colored, without a reddish tinge; the underside is gray to gray white. The fur is is soft and sparse. Juveniles are dark gray.

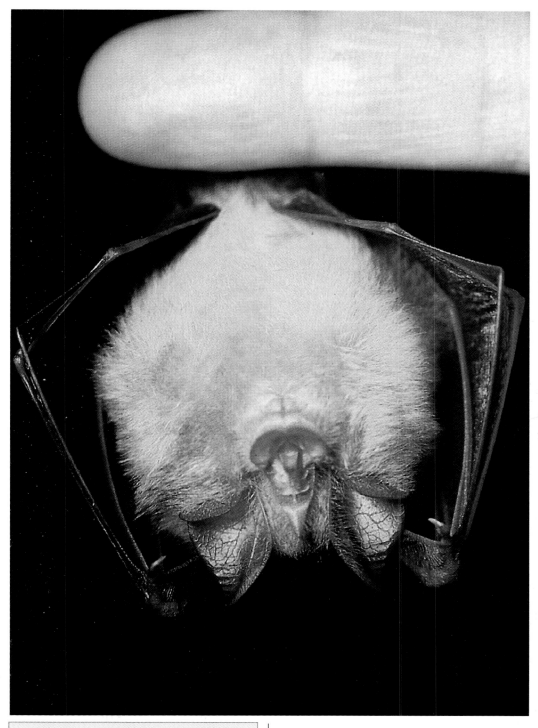

The Lesser Horseshoe Bat is one of the smallest and most delicate European bat species.

The ears and wing membranes are light gray-brown. It wraps itself completely inside its wing membranes in hibernation and at rest. False teats are fully developed in females in the second year.

Anomalies in coloration: Flavism (pigment deficiency with yellow coloration).

Similar species: Unmistakable on account of the small body size and rounded upper saddle process.

Range: This is the horseshoe bat with the northernmost distribution, reaching a latitude of approximately 52° north. It occurs in western Ireland and southwestern England, France, Belgium, southern Netherlands, Luxembourg, western Germany (northernmost locality in lower Mosel), eastern Germany (southern Middle and Lower Harz, northern Thuringia, Dresden basin), southern Poland, Czech Republic, Slovakia, Ukraine, and Russia (Caucasus).

Biotope: It inhabits the warmer regions in foothills and sub-alpine mountains, in part in forested landscapes, and in karst (limestone) formations. In summer it has been documented up to an altitude of 1160 meters, in winter up to 2000 meters altitude, with the highest maternity roost occurring at 950 meters. It is a house bat in the north and a cave bat in the south. Summer roosts (maternity roosts) in the north are in warm attics, often near the chimney, in drains, and in shafts of heated basements; in the south it roosts in caves and tunnels. The roost must always be free of drafts.

It roosts in winter in caves,

tunnels, and cellars, the temperature being 6 to 9°C, with high humidity. In the winter, it roosts always free-hanging while maintaining a distance from its neighbors, with up to 300 individuals in the roost. It hibernates from September or October to late April. Males often predominate in the winter roosts; they also occupy them before the females. Hanging height is from just above ground level up to 20 meters high.

Migrations: A permanent resident. It only migrates 5 to 10 kilometers between summer and winter roosts, the longest migration being 153 kilometers.

Reproduction: Females are sexually mature in their first year (in Slovakia approximately 15% of the females bear young in their first year).

Mating takes place in autumn, partly in the winter roost. Before mating, pairs may court by chasing. The male then hangs behind and over the female. The duration of copulation is very brief.

Maternity roosts are in part combined with other bats (Greater Mouse-eared Bat, Geoffroy's Bat), though not mingled directly together with other species. Maternity roosts are occupied beginning in about April by about 10 to 100 females, along with some males (up to 20%).

Starting in mid June to early July, about one-half to two-thirds of the females present in the maternity roost give birth to a single youngster. At birth the young weigh about 1.8 grams and

Lesser Horseshoe Bat.

have a forearm length of approximately 15 to 19 millimeters. They are covered with fine hair, except on the underside, and have sensory hairs in the horseshoe area. The eyes open on about the tenth day, and the young are completely independent at 6 to 7 weeks. The maternity roosts disband in August.

Maximum age: 21 years; average age 3 to 4 years.

Hunting and diet: Their flight is agile and relatively fast, the wing movements almost whirring. They hunt in open forests and parks, among undergrowth. The flight altitude is low, up to 5 meters. they also take up prey from rocks and branches.

They prey on small moths, mosquitoes, crane flies, beetles, and spiders.

Calls: Chirping or scolding.

The echolocation call is a sustained, constant frequency tone of 105 to 111 kHz, ending in a short drop in frequency. The call duration is approximately 20 to 30 ms. The call of this species overlaps in frequency with the Mediterranean Horseshoe Bat and Mehely's Horseshoe Bat.

Protection: Threatened with extirpation in both western and eastern Germany and endangered in Austria. It has experienced a drastic decline in population throughout central Europe, and has already disappeared in northern central Europe. The causes include the disturbance and destruction of roosts and the use of insecticides, and in the northern part of the range possibly also a worsening in climate. The systematic protection of summer and winter roosts is necessary! Keep entrance openings free (approximately 15 x 25 centimeters)!

RHINOLOPHUS FERRUMEQUINUM (SCHREBER, 1774)

Eng.: Greater Horseshoe Bat
Ger.: *Grosse Hufeisennase*
Fr.: *Grand rhinolophe fer á cheval*
Head-body: (50) 57-71 mm
Tail: (30) 35-43 mm
Forearm: (50) 54-61 mm
Ear: 20-26 mm
Wingspan: 350-400 mm
Condylobasal length: (19) 20-22 mm
Weight: 17-34 g

Distinguishing characters: The largest European horseshoe bat. The upper saddle process is short and rounded, the lower process tapering to a point in profile. The fur is soft and sparse The hair is light gray at the base, the back gray brown or smoky gray, more or less with a reddish tinge. The underside is gray-white to yellowish white. Juveniles are more ashy gray on the back. The wing membranes and ears are light gray-brown.

It wraps itself completely inside its wing membranes during hibernation and usually also in daytime lethargy. As with the Lesser Horseshoe Bat, it ends the third to fifth finger in the basal joint and the joint between the first and second phalanges only slightly. In females the false teats are not fully developed until the third year.

Anomalies in coloration: Unknown.

Similar species: This species is unmistakable because of its size and the blunt upper saddle process.

Range: Central and southern Europe. In the north it occurs approximately to 51°41' latitude, also in southwestern England (Devonshire), France, southeastern Belgium, Netherlands (southern Limburg), Luxembourg, western Germany (region of Kreuznach, Kaiserstuhl-Freiberg, southern

Greater Horseshoe Bat.

Franconian Alps; no reports for eastern Germany), southern Poland (only one report), southeastern Slovakia, in the east to the Caucasus. In the south it is documented in all the Balkan and Mediterranean countries.

Biotope: Occurs in warmer regions with open stands of trees and shrubs, standing or flowing bodies of water, and karst formations, as well as in towns. It is more of a house bat in the north and a cave bat in the south. In mountains it is usually found at altitudes below 800 meters, only rarely to 2000 meters. In the north summer roosts (maternity roosts) are usually in warm attics and church steeples; in the south they roost primarily in caves and tunnels.

Winter roosts are located in caves and tunnels with temperatures between 7 and 10°C, rarely lower. It hangs freely from the ceiling, rarely forming small clusters (for thermoregulation).

Hibernates from September or October to April, but can be interrupted once or twice a week. In mild weather it may also hunt for food near the cave entrance.

Migrations: A permanent resident. The distance between summer and winter roosts is, as a rule, 20 to 30 kilometers, with the longest migration 180 kilometers.

Reproduction: Females are not usually sexually mature until they are three years of age (England), or two years on the European mainland; males are sexually mature at the earliest at the end of their second year.

The breeding season extends from fall to spring. Maternity roosts contain up to 200 females. Males are also present in the maternity roosts. In the maternity roosts the females hang with their young both singly and in clusters. They at times associated with the Mediterranean Horseshoe Bat and Geoffroy's Bat.

Females give birth to one youngster, which opens its eyes at about 7 days, starting from about mid June to July. Juveniles are

Greater Horseshoe Bat.

ready for flight at 3 to 4 weeks of age, and are independent at 7 to 8 weeks (mid August).

Maximum age: 30 years. This is the greatest age attained so far by any European bat.

Hunting and diet: This bat emerges from its roost at the onset of darkness. Its flight is low, usually at low altitude (0.3 to 6 meters), butterfly-like, with short sections of gliding. There is little flight activity in cold, windy, and rainy weather.

It hunts in the country where there are open stands of trees, on slopes, rock faces, as well as in gardens. It can locate and then catch insects from perches ("flycatcher bat"); it also takes prey from the ground.

It preys on larger insects (June beetles, carrion beetles, grasshoppers, moths) and uses feeding roosts.

Drinking is accomplished during low flight or while hovering. The active radius of colonies in England is 8 to 16 kilometers.

Calls: Emits a relatively low, chirping or chattering call.

The echolocation call is a sustained constant frequency tone of 77 to 81 kHz, ending in a short drop in frequency (see the sonogram). The call duration is about 30 to 40 ms.

Protection: This species is threatened with extirpation in western Germany and is severely threatened in Austria.

As all horseshoe bats, the Greater Horseshoe Bat is sensitive to disturbances. Therefore, systematic protection of maternity roosts and winter roosts (free entrance openings of about 20 x 30 centimeters) is required, as well as preservation of the food supply (no use of insecticides, danger of extermination of large beetles through changes in soil preparation in agriculture). There is a general severe population decline in Central Europe.

RHINOLOPHUS EURYALE BLASIUS, 1853

Eng.: Mediterranean Horseshoe Bat

Ger.: *Mittelmeerhufeisennase*

Fr.: *Rhinolophe euryale*

Head-body: 43-58 mm

Tail: 22-30 mm

Forearm: 43-51 mm

Ear: 18-24 mm

Wingspan: 300-320 mm

First phalanx, fourth finger: 6.6-8.5 mm

Second phalanx, fourth finger: 17.9-19.1 mm

Condylobasal length: 16-17.6 mm

Weight: 8-17.5 g

Distinguishing characters: Medium sized. The upper saddle process is pointed and slightly downward curving, and is distinctly longer than the lower saddle process, which is broadly rounded when viewed from below. The bare part of the face (horseshoe, lips) is light brownish; the ears and wing membranes are light gray. The fur is sparse and light gray at the base. The back is gray-brown, with a light reddish or pinkish tinge. The underside is gray-white to yellowish white; the boundary between the back and underside is indistinct; often a few darker hairs are present around the eyes. Juveniles are collectively more gray colored. The wings are broad, the second phalanx of the fourth finger is more than twice as long as the first phalanx; the third to fifth fingers at rest are bent 180° at the joint between the first and second phalanges; therefore, as a rule (even in hibernation) they are not completely wrapped in the wing membranes.

In the active state, Mediterranean Horseshoe Bats often maintain body contact while hanging, embracing one another with their wings and licking one another's face and head.

Anomalies in coloration:

Mediterranean Horseshoe Bat echolocating just prior to take-off. The mouth is closed because the echolocation calls are sent out through the nostrils.

Albinism.

Similar species: In Mehely's and Blasius's Horseshoe Bats, pay attention to the shape of the nasal processes, the length of the first and second phalanges of the fourth finger, and the fur coloration; in the Greater and Lesser Horseshoe Bats, observe the differences in size and the different shape of the saddle.

Range: Occurs in the Balkans and Mediterranean region, as well as on Sicily, Sardinia, and Corsica. Their northernmost occurrence is in Slovakia, northern Italy, and southern France.

Biotope: Inhabits warm, forested regions in foothills and mountains; prefers karst formations with numerous caves and water nearby.

Summer roosts (maternity

Mediterranean Horseshoe Bat.

roosts) are located in caves. Also, rarely, in northern regions in warm attics (cave bat). Maternity roosts contain 50 to 400 females, with males often present.

Often roosts together with other horseshoe bats, Geoffroy's Bat, and Schreiber's Bat. Winter roosts are located in caves and tunnels, in which the temperature is approximately 10°C. Hangs freely from the ceiling, to some extent while maintaining body contact with other conspecifics.

Migrations: Generally a permanent resident; longest migration 134 kilometers.

Reproduction: No detailed information available.

Females give birth to one youngster, weighing about 4 grams at birth, that are ready for flight in early to mid August. In Bulgaria, the young are ready for flight starting in mid July; at the same time pregnant females are still present.

Maximum age: Unknown.

Hunting and diet: The Mediterranean Horseshoe Bat emerges from its roost in late twilight. It hunts at low altitude on warm slopes, but also in relatively dense stands of trees or shrubs. Its flight is slow, fluttering, and very agile, and it can hover.

It preys on moths and other insects, eating its prey to some extent at the feeding roosts.

Protection: Populations are declining in the northern part of its range, particularly in France and Slovakia. A systematic protection of the roosts and biotope is necessary!

Calls: Low chirping, squeaking, or scolding.

The echolocation call is a constant frequency call from 101 to 108 kHz, ending in a short drop in frequency. The call duration is approximately 20 to 30 ms. The frequency overlaps with the Lesser Horseshoe Bat and Mehely's Horseshoe Bat.

RHINOLOPHUS BLASII (PETERS, 1866)

Eng.: Blasius's Horseshoe Bat
Ger.: *Blasius Hufeisennase*
Fr.: *Rhinolophe de Blasius*
Head-body: (44) 46.5-54 (56) mm
Tail: (20) 25-30 mm
Forearm: (43.5) 45-48 mm
Ear: 16.5-21 mm
Wingspan: approximately 280 mm
First phalanx, fourth finger: approximately 8.2 mm
Second phalanx, fourth finger: approximately 14-15 mm
Condylobasal length: 15.8-16.7 mm
Weight: (10) 12-15 g

Distinguishing characters: Medium sized. The upper saddle process is pointed, straight, not curving downward, and longer than the lower saddle process, which is narrowly rounded when viewed from the front. The vertical furrow is slightly notched in the middle and the lancet tapers uniformly upward. The horseshoe is broad, flesh colored, and the ears and wing membranes are light gray. The fur is sparse, very light colored, nearly white, at the base. The

Blasius's Horseshoe Bat.

back is gray-brown, in part with a light pink tinge. The underside is nearly white or has a slightly yellowish tinge. The boundary between the back and underside is relatively sharp. Dark "spectacles" around the eyes are absent or only suggested. The wings are broad, and the second phalanx of the fourth finger is, at most, twice as long as the first phalanx.

Anomalies in coloration: Unknown.

Similar species: Note the length of the first and second phalanges of the fourth finger and the shape of the nasal processes in contrast with the Mediterranean and Mehely's Horseshoe Bats. Note the difference in size and different saddle shape between this species and the Greater and Lesser Horseshoe Bats.

Range: Only sketchy information is available. It occurs in northern Italy, Greece, Croatia, Bosnia, Montenegro, Albania, Bulgaria (Black Sea coast), Sicily, and the Caucasus.

Biotope: Warm karst formations with open stands of shrubs and trees are preferred.

Summer and winter roosts are in caves (cave bat).

No precise information is available on hibernation. They are free hanging, and there is no mutual body contact.

Protection: No reliable information on the degree of

endangerment. Protection of the roosts necessary.

Migrations: Unknown, this species is probably a permanent resident.

Reproduction: Maternity roosts are located in caves and contain up to 200 females. Females give birth to one youngster. Otherwise no detailed information is available.

Maximum age: Unknown.

Hunting and diet: Probably similar to the Mediterranean Horseshoe Bat.

Calls: The echolocation call is a constant frequency tone of 93 to 98 kHz, ending in a short frequency drop (see sonogram). The call duration is approximately 40 to 50 ms.

RHINOLOPHUS MEHELYI (MATSCHIE, 1901)

Eng.: Mehely's Horseshoe Bat
Ger.: *Mehely Hufeisennase*
Fr.: *Rhinolophe de Mehely*
Head-body: (49) 55-64 mm
Tail: (23) 24-29 (32) mm
Forearm: (47) 50-55 mm
Ear: 18-23 mm
Wingspan: approximately 330-340 mm
First phalanx, fourth finger: 7.7 mm
Second phalanx, fourth finger: 19 mm
Condylobasal length: (16.1) 16.6-17.5 (18) mm
Weight: 10-18 g

Distinguishing characters: Medium sized. The upper saddle process is relatively blunt in profile, and only slightly longer than the lower process, which is broadly rounded when viewed from the front. The lancet narrows abruptly upward and tapers to a thin point. The horseshoe and lips are pale, flesh colored, the ears and wing membranes gray-brown. The fur is relatively dense and gray-white at the base. The back gray-brown, the underside nearly white; the boundary between the back and underside is relatively sharp. Conspicuous dark spectacles of gray-brown hairs are present around the eyes. The wings are broad. The second phalanx of the fourth finger is more than twice as long as the first phalanx. At rest the third to fifth fingers are bent at the joint between the first and second phalanges by 180°; therefore, they are not completely wrapped when hanging.

Anomalies in coloration: Unknown.

Similar species: When comparing it with the Mediterranean Horseshoe Bat particularly note the different shape of the lancet and the upper saddle process; compared with Blasius's Horseshoe Bat pay attention to the length of the first and second phalanges of the fourth finger and the nasal processes. Compared with the Greater and Lesser Horseshoe Bats, take into account the shape of the saddle and differences in size.

Range: Only sketchy information is available. It is found in southern France (Mediterranean coast), southern Italy, Sicily, Corsica, Sardinia, Greece, Croatia, Bosnia,

Mehely's Horseshoe Bat in flight. Because of the short tail, the tail membrane of the horseshoe bat in flight always forms a constant curve. In the common bats, on the other hand, it is pointed or convex.

Montenegro, Serbia, Bulgaria, Romania (Dobrudscha), and the Caucasus.

Biotope: A cave bat. As far as is known, winter roosts are located in caves and in karst formations with water nearby. Occurs partly together with other horseshoe bats, the Lesser Mouse-eared Bat, and Schreiber's Bat. They roost hanging free from the ceiling. No detailed information is available on its hibernation.

Migrations: Unknown, probably a permanent resident.

Reproduction: No detailed information is available. Maternity roosts contain up to 500 individuals (Romania).

Females give birth to one youngster. Juveniles in Romania are capable of flight by the second half of July.

Maximum age: Unknown.

Hunting and diet: This species emerges from the roost at dusk.

Hunts low over the ground on warm mountain slopes, also among shrubs and trees. The flight is slow, but very nimble and agile, is interrupted by short stretches of gliding, and can take

Mehely's Horseshoe Bat. Note the dark "spectacles" in the eye region.

off from the ground effortlessly (also takes prey from the ground?).

Calls: Relatively low, loud, short chirping or peeping.

The echolocation call is a constant frequency tone of 105 to 112 kHz, ending with a short drop in frequency. The call duration is approximately 20 to 30 ms. There is an overlap in frequency between this species and the Lesser Horseshoe and the Mediterranean Horseshoe Bats.

Protection: There is no reliable information on the degree of endangerment. Protection of the roosts and biotope is necessary.

Common Bats, Family Vespertilionidae

There are 40 genera with about 320 species, 24 of which occur in Europe. The muzzle is smooth, without nasal processes, the ear has a tragus, and the eyes are usually small. The tail is completely (or except for the last two vertebrae) included in the tail membrane, and folded under the belly at rest. The wings vary from long and narrow to broad and short. At rest they are folded together along the sides of the body, with the third to fifth finger rotated 180° to the inside around the basal joint (exception: Schreiber's Bat). The feet are equipped with a spur, in part with epiblema.

The fur color is inconspicuous (black, brown, or gray), with the belly lighter than the back.

Two milk teats are present (one exception, the Parti-colored Bat), while false teats are absent. Milk teeth are present at birth. As far as the permanent teeth are concerned, the number of premolars varies from genus to genus:

$$\text{tooth formula } \frac{211\text{-}33}{312\text{-}33} = 32\text{-}38$$

In some species flight is fast and persistent. Long migrations between summer and winter roosts are sometimes undertaken.

Prey are usually captured in flight. Several species also take prey on the ground or from foliage, branches, or other substrates. Cold-hardy species reach the Arctic Circle. Echolocation is accomplished with frequency modulated (FM) calls. Many species can be distinguished on the basis of the specific frequency of their echolocation calls. Calls are sent out through the open mouth (exception: the Gray Long-eared Bat also sends calls through the nose).

Genus: *Myotis* Kaup, 1829

Myotis includes approximately 90 species, 10 of which occur in Europe. The ears are longer than wide, the tragus is usually long and lanceolate, and the spur lacks epiblema, at most there is a narrow edging of skin suggested. Very large to small species are included. There is one pair of teats.

tooth formula	$\dfrac{2\ 11\text{-}33}{3\ 12\text{-}33}$ =32-38

MYOTIS DAUBENTONI (KUHL, 1819)

Eng.: Daubenton's Bat
Ger.: *Wasserfledermaus*
Fr.: *Vespertilion de Daubenton*
Head-body: (40) 45-55 (60) mm
Tail: (27) 31-44.5 (48) mm
Forearm: (33) 35-41.7 (42) mm
Ear: 10.5-14.2 mm
Wingspan: 240-275 mm
Condylobasal length: 13.2-14.6 mm
Weight: (5) 7-15 g

Distinguishing characters:
Medium-sized to small species. The outermost ear margin has a slight indentation in the lower half. The ear is relatively short, with four to five horizontal creases. The tragus is straight, tapers upward, and does not reach half of the ear length. When disturbed, the ear is oriented nearly at right angles to the side. The fur is sparse and gray-brown at the base. The back is brown-gray to dark bronze colored, the tips of hairs often shiny. The underside is silvery gray, in part with a brownish tinge. The boundary between the back and underside is usually sharp. The muzzle is rufous, with the ears and wing membranes dark gray-brown. Juveniles are more gray, darker. The feet are large and have long bristles. The spur attains a length of about one-third of the length of the tail membrane; at three-quarters of the length of the spur, however, there is a distinct break in the

Daubenton's Bat.

margin of the tail membrane, acting as a spur tip. The arm membrane attaches at the base of the toe.

Anomalies in coloration: Albinism has repeatedly been observed.

Similar species: Pond Bat: Larger. An overlap in size is possible with a number of other species of *Myotis*.

Long-fingered Bat: The tail membrane is hairy above, the fur is more gray, and the wing membrane always attaches above the ankle.

Natterer's Bat: The spur is curved, S-shaped. Fringing hairs are present on the margin of the tail membrane. The ears and tragus are lighter and longer.

Geoffroy's Bat: The fur is usually conspicuously rufous, and the outer margin of the ear is more clearly indented.

Bechstein's Bat: The ears are distinctly longer.

> The open mouth (echolocating) and the already raised left wing show that this Daubenton's Bat is in the act of taking off.

Brandt's Bat: The tragus is long, projecting beyond the indentation on the hind margin of the ear. The spur attains at most half the length of the tail membrane; there is no break in the tail membrane at three-quarters of the length the length.

The Lesser Daubenton's Bat (*Myotis nathalinae*) was described in 1977 by Tupinier from Spain and France, with additional records from Switzerland and Poland.

This species is very similar to Daubenton's Bat. Its back is more gray. In contrast to Daubenton's Bat, it lacks the protoconus on the inside of the fourth premolar. The baculum (penile bone) differs in shape from that of Daubenton's Bat.

Studies of Daubenton's Bat from Poland, eastern German, the Czech Republic, and Slovakia, yielded, in part, similar premolar forms as were described for *M. nathalinae*, but not in combination with lower body weights and a different form of the baculum.

Based on more recent studies, however, *M. nathalinae* is not an independent species, but rather only a morphotype within the spectrum of variation of Daubenton's Bat.

Range: This species occurs throughout almost all of Europe. In the north, it extends nearly to 63° latitude, and is absent only in northern Scandinavia and northern Scotland. In the south, there are individual records in the Balkan countries (Romania, Bulgaria, Croatia, Bosnia, Serbia, and Montenegro). Apparently it barely overlaps the range of the Long-fingered Bat.

Biotope: Lives primarily in lowlands, forests, and parks, usually near water. It is a forest bat. In summer it is found up to 750 meter altitude, in winter it has been documented up to 1400 meters.

Summer roosts (maternity roosts) are located in tree holes. The round or narrow entrance holes are sometimes below one meter. It is also found in attics. Individuals and small groups of males often roost in cracks under bridges and in walls, rarely in bat boxes.

Winter roosts are located in caves, tunnels, bunkers, cellars, and old wells. The temperature there is 3-6°C (8°C), but may drop temporarily to -2°C; high humidity. Usually they are pressed into cracks, but are also free hanging in large clusters on the wall. There may sometimes be up to 100 individuals on and over one another like roof tiles. Has also been found up to 60 centimeters deep in the floor rubble. Large winter roosts contain more than 1000 individuals.

Hibernation occurs from late September or mid October to late March or April. The females arrive in the winter roost before the males. Invasion-like flights to future winter roosts have been observed in August.

Migrations: A partial migrant, usually traveling less than 100 kilometers. It flies to its winter roost from all directions. The longest migration recorded is 240 kilometers.

Reproduction: Some females apparently become sexually mature at one year of age.

The mating season extends from September to spring, with breeding often occurring in the winter roost. Beginning in May, maternity roosts are occupied by 20 to 50 females, rarely up to 200. During this time males gather in groups of up to 20 individuals; individual males also occupy maternity roosts.

The young are born starting in the second half of June. At birth the back, ears, and wing membranes are gray-brown, the underside pink. Very fine, short hairs are already present on the back, and sensory hairs are present on the tail. At birth they weigh approximately 2.3 grams, and the forearm is 24.1 millimeters. After ten days the weight is 4.3 grams, the forearm is 24.1 millimeters. At 21 days of age the weight is 5.5 grams, and the forearm is 32.7 millimeters. The eyes open after 8 to 10 days. The coat is complete starting on the twenty-first day, the hair growth is finished at 31 to 55 days. Permanent teeth are complete at about the thirty-first day. Juveniles are capable of flight in the fourth week. The maternity roosts disband in August.

Protection: Although this species is endangered in western Germany and Austria, it is only threatened locally in eastern Germany. According to observations in winter roosts in certain countries in Central Europe, there has been a more or less clear increase in population. Possible reasons can be found in the hunting habits or in the dietary spectrum (no insecticide use over water). Daubenton's Bats are occasionally caught by fishermen (hooks in the mouth or wings). Protection is needed through preservation of roosts (tree holes) and biotopes!

Maximum age: 20 years. The average age is 4 to $4^1/_2$ years.

Hunting and diet: It emerges from its roost at dusk. The flight is fast, agile, with fast, in part whirring, wing beats.

Often it hunts only 5 to 20 centimeters above the water's surface, but also up to 5 meters high around trees. During breaks from hunting it hangs on branches or walls.

It preys on small flying insects (mosquitoes, crane flies, moths, and the like), feeding in flight.

The hunting grounds are usually located only 2 to 5 kilometers from the roost.

Calls: This species produces chirping calls in flight, a shrill scolding in defense, and a persistent, shrill call when disturbed in hibernation.

Echolocation calls (searching flight) are FM calls from 69 to 25 kHz (78 to 32 kHz) with a duration of 3 to 4 ms. The call

Various morphotypes of the fourth premolar in the upper mandible (P4) in a series of Daubenton's Bats. The development of the fourth premolar in the right row (well-developed protoconus on the inside) corresponds to the form that Tupinier (1977) described for *Myotis nathalinae* (from Hank, 1983/84).

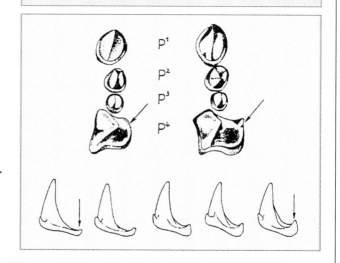

exhibits a sinus-like amplitude modulation (approximately 10 maxima) with the greatest intensity at 45 kHz. Calls are repeated at (35 to) 75 ms intervals, approximately 13 (to 28) calls per second. The range is 20 to 40 meters (see sonogram).

MYOTIS CAPACCINII (BONAPARTE, 1837)

    Eng.: Long-fingered Bat
    Ger.: *Langfussfledermaus*
    Fr.: *Vespertilion de Capaccini*
    Head-body: (43) 47-53 mm
    Tail: 35-42 mm
    Forearm: (37) 38-44 mm
    Ear: (13) 14-16 mm
    Wingspan: 230-260 mm
    Condylobasal length: 13.9-14.8 mm
    Foot: 11-13 mm
    Weight: 6-15 g

Distinguishing characters: Medium sized. The ears are of medium length and narrow, with the outside margin slightly indented toward the tip and with five horizontal creases. The tragus is pointed, reaching half the length of the ear. The inside margin is convex, the outside margin is concave, arched, and has a suggestion of serrations. The fur is dark gray at the base. The back is light smoky gray, partially with a slight yellowish tinge, the underside light gray. The boundary between the back and the underside is indistinct. The muzzle is reddish brown, the ears and wing membranes gray brown. The feet are conspicuously large and have long bristles. The tail membrane is covered with dense, dark, downy hair above and below from the legs to about the middle. Hairs extend beyond the margin of the tail membrane in the spur region. The spur is straight, attaining about one-third the length of the margin of the tail fin; at two-thirds to three-quarters the length there is a break in the margin of the tail membrane that acts as a spur end. The arm membrane is broad, attaching on the lower leg 3 to 5 millimeters above the

Long-fingered Bat.

ankle. The nostrils are clearly more protruding than in other European *Myotis* species.

Anomalies in coloration: Unknown.

Similar species: In Daubenton's, Natterer's, Pond, Bechstein's, and Geoffroy's Bats, the dorsal side of the tail membrane is not hairy. For additional characters, see Daubenton's Bat and the determination keys.

Range: Occurs in the European Mediterranean region and the Balkan countries, but only sketchy information is available. The northern limit of the range is in Spain, southern France, Italy, Switzerland (southern Tessa), Bulgaria, Croatia, Bosnia, Serbia, Montenegro, Bulgaria, and Greece. Apparently the range of this species only slightly overlaps the range of Daubenton's Bat.

Biotope: Inhabits karst formations and forested and shrubby landscapes near water.

The summer and winter roosts are located in caves (cave bat); in the winter roost often occurring in crevices.

Migrations: Unknown. Apparently this species is a permanent resident or partial migrant.

Protection: Because little is known about the habits of this species, the degree of endangerment cannot be determined. Protection of the caves (summer and winter roosts) is necessary.

Reproduction: Only sketchy information is available. Maternity roosts are located in caves with up to 500 females in clusters on the cave ceiling.

Maximum age: Unknown.

Hunting and diet: It emerges from its roost in late evening. The flight resembles that of

Long-fingered Bat.

Daubenton's Bat. This bat often hunts over water.

Calls: Audible calls are a shrill scolding, similar to Daubenton's Bat.

MYOTIS DASCYNEME (BOIE, 1825)

Eng.: Pond Bat
Ger.: *Teichfledermaus*
Fr.: *Vespertilion des marais*
Head-body: 57-67 (68) mm
Tail: (39) 46-51 (53) mm
Forearm: (41) 43-49.2 mm
Ear: (14.9) 16-19 mm
Wingspan: 200-300 mm
Condylobasal length: 15.7-17.4 mm
Weight: (11) 14-20 (23) g

Distinguishing characters: Medium sized. The outside ear margin lacks a distinct indentation, but has five horizontal creases. The tragus is distinctly shorter than half the ear length, very short for a *Myotis* species, narrowing only slightly toward the tip, and rounded to a point, curving slightly inward. The fur is dense, black brown at base. The back is brownish or pale gray-brown with a silky sheen. The underside is gray-white to yellowish gray, and set off relatively sharply from the back. Juveniles are altogether darker in color. The short muzzle is rufous, the ears and wing membranes gray brown. The wings are long and broad, with the arm membrane attaching at the ankle. The feet are large and have long bristles. The underside of the tail membrane along the lower leg has fine white hairs. Hairs project beyond the margin of the tail membrane at the spur.

The spur is straight, reaching about one-third of the length of the tail fin. At three-quarters of the length there is a distinct break in the margin of the tail fin, which acts as a spur tip.

Anomalies in coloration: Unknown.

Similar species: The very similarly colored Daubenton's Bat is smaller. In addition compare the tail membrane and tragus.

The Long-fingered Bat has a hairy dorsal side of the tail membrane. For additional details, see the diagnostic key.

Range: This species occurs in a broad band between 48 and 60° latitude in central and eastern Europe. It is found from northeastern France through Belgium, Holland, Denmark, southern Sweden, and Poland to Lithuania, Latvia, and Estonia. In the south it occurs throughout the Czech Republic, Slovakia, and Hungary to the Ukraine and Belarus. In eastern and western Germany, it is found mostly only sporadically in winter roosts; in eastern Germany it is also found sporadically in the summer. The centers of distribution with maternity roosts are located in the Netherlands (Friesland and northern Holland), Denmark (Jutland), and Lithuania.

Biotope: In the summer it inhabits areas with abundant water, meadows, and forests in lowlands, and in the winter also in foothills of low mountains. It has been documented up to an altitude of 1000 meters. Winter roosts, as a rule, are scarcely above 300 meters elevation. Summer roosts (maternity roosts) are usually found in large groups in attics or church steeples, often on darker sections of the roof ridge. Individuals also roost in hollow trees.

Winter roosts are located in natural caves, tunnels in karst

up to several hundred individuals are present, which may also form small clusters.

Migrations: This species is a partial migrant. It migrates from the northern summer roosts to the southern winter roosts, as a rule traveling more than 100 kilometers. The longest migration

Pond Bat.

formations, cellars, and bunkers, where there is a temperature range of 0.5 to 7.5°C.

Hibernation occurs from October to mid March or April. It hibernates both wedged into cracks and free hanging on walls or ceilings. In large winter roosts

recorded so far was 330 kilometers.

Reproduction: Females probably become sexually mature in their second year.

The mating season begins in late August. Mating also takes place in the winter roost. Maternity roosts are occupied in May by 40 to 400 females, rarely by males. During this time males live singly or in small groups, distributed over a wide area. Females can change the maternity roosts from year to year.

Starting in mid June females give birth to a single youngster, which becomes independent starting in about mid July.

water, but also over meadows and along the edges of forests. The flight is fast, agile, and often only 5 to 10 centimeters above the water. It preys on mosquitoes, crane flies, and moths, and also takes insects from the water's surface.

Calls: The echolocation call (searching flight) is an FM call

Pond Bat. Note the large feet.

Maternity roosts break up in August.

Maximum age: 19 years.

Hunting and diet: This bat emerges from the roost in late evening.

There are two hunting times, in the evening and towards morning. It not only hunts over

from 60 to 24 kHz, in part merging into a CF component. The duration of the call is 5 to 8 ms, the maximum impulse intensity 36 to 40 kHz. Calls are repeated at 115 (90 to 130) ms intervals (approximately 8 to 10 calls per second); the range is 5 to 20 meters (see the sonogram).

Protection: Populations are declining in the western part of the range. In western Germany it is an endangered migratory animal, in eastern Germany it is a rare species. Systematic protection of the maternity roosts and winter roosts is necessary in the few known centers of occurrence. Protection of the biotope is mandatory. Exercise caution with wood preservatives!

MYOTIS BRANDTI (EVERSMANN, 1845)

Eng.: Brandt's Bat
Ger.: *Grosse Bartfledermaus*
Fr.: *Vespertilion de Brandt*
Head-body: 39-51 mm
Tail: 32-44 mm
Forearm: 33-39.2 mm
Ear: 13-15.5 (17) mm
Wingspan: 190-240 mm
Condylobasal length: 13.1-14.4 mm
Weight: 4.3-9.5 g

Its occurrence in Europe was first discovered in 1958 by Topal.

Distinguishing characters: The outside margin of the ear has distinct indentation, which the long, pointed tragus projects beyond. Four or five horizontal creases are present.

The fur is relatively long, the hair dark gray-brown at the base. The back is light brown, usually with a golden sheen. The underside is light gray, partially with a yellowish tinge. the muzzle, ears, and wing membranes are medium to light brown. The base of the tragus and the inside of the ear margin are clearly lighter. The wings are relatively narrow, with the arm membrane attaching at the base of the toe. The feet are small. The spur is shorter than half the length of the tail membrane; a narrow edging of skin is usually present on the outside margin. The penis in adult males is distinctly thickened at the tip. The cusp next to the third premolar is higher or as high as the cusp alongside the second premolar; the second premolar is not clearly smaller than the first premolar. Juveniles closely resemble the Whiskered Bat. They have the back dark gray-black to gray-brown, and the muzzle and ears are black brown. This species is lively in behavior, but not as temperamental as the Whiskered Bat.

Anomalies in coloration: Unknown.

Similar species: In the Whiskered Bat, the males lack the thickening at the tip of the penis. The color of the ears and muzzle is black-brown, while the base of the tragus and the inside margin of the ear are not lighter in color.

Daubenton's Bat has a shorter tragus, and take special note of the length and shape of the spurs and tail membrane.

Geoffroy's Bat generally has a rufous back, and never has a golden sheen. The ear shape and tragus length are important also.

Range: Only sketchy information is available, because formerly this species was not differentiated from the Whiskered Bat. It occurs in northern England (except for Scotland), France, Belgium, the Netherlands, eastern and western Germany, in

Brandt's Bat with nearly full-grown juvenile. In this bat species the difference in color between juveniles and adults is particularly great.

Scandinavia approximately to 64° latitude, Poland, Latvia, Lithuania, and Estonia. In the south it extends to Spain, Italy, and the Balkan countries. It is widely distributed in Asia.

Biotope: This is a forest bat. It is more closely bound to forests and bodies of water and less to human settlements than is the Whiskered Bat. In winter it has been documented up to an altitude of 1730 meters; the highest maternity roost is 1270 meters (Switzerland).

Summer roosts (maternity roosts) are found in narrow cracks in the attics of buildings, behind roof laths, in holes in beams, and in shallow bat boxes.

Winter roosts are in caves, tunnels, old mines, and cellars; temperatures are (0°C) 3-4°C (7.5°C). It often roosts together with Whiskered Bats, usually free hanging on the wall or ceiling, rarely in cracks, as well as together in clusters with Daubenton's Bat. It hibernates approximately from October to March or April.

Migrations: This is a partial migrant. Its longest migration was 230 kilometers.

Reproduction: The onset of sexual maturity in females is not precisely known, but it probably occurs in the second year.

Mating takes place in the fall and in the winter roost. Maternity roosts are occupied in May at the latest by about 20 to 60 females. Mixed maternity roosts with Nathusius's

Pipistrelle are possible in shallow boxes.

Females give birth to one youngster from mid June to early July. Newborns are dark gray on the back with a lighter underside. The ears are limp and the back is covered with barely visible, fine hair. At 10 days the weight is 3 grams and the forearm is 20.3 millimeters. At 22 days the weight is 4.5 grams, the forearm is 32.2 millimeters, and the wingspan is 200 millimeters. The eyes open on the third day and the ears become erect between the fifth and ninth days.

Brandt's Bat.

Young are capable of flight at about three to four weeks of age. The penis is not yet thickened at the tip in young males.

Maximum age: 19 years, 8 months.

Hunting and diet: It emerges from the roost in early evening.

It hunts at low to medium altitude in not-too-dense forest, often over water. The flight is fast and agile with quick turns, but it is not as agile as the Whiskered Bat in confined spaces.

The dietary spectrum not precisely known. It probably consists of small moths and other flying insects.

Calls: When disturbed, it emits high-pitched scolding and

chirping. Small young emit a persistent, very high-pitched twittering (contact call).

The echolocation (searching flight) call behavior is the same as that of the Whiskered Bat.

Protection: This species is severely endangered in western Germany, endangered in Austria, and a rare animal in eastern Germany. Putting out shallow bat boxes in suitable biotopes is highly recommended!

MYOTIS MYSTACINUS (KUHL, 1819)
    Eng.: Whiskered Bat
    Ger.: *Kleine Bartfledermaus*
    Fr.: *Vespertilion á moustaches*
    Head-body: 35-48 mm
    Tail: 30-43 mm
    Forearm: (31) 32-36 (37.7) mm
    Ear: 12-17 mm
    Wingspan: 190-225 mm
    Condylobasal length: 12.3-13.3 (13.6) mm
    Weight: (3) 4-8 g

Distinguishing characters: Smallest European *Myotis* species. The outside ear margin has a distinct indentation, the tragus is projecting, long, and pointed, and there are four or five horizontal creases. The muzzle, ears, and wing membranes are blackish brown. The base of the tragus and the inner ear margin are not lighter in color in contrast to Brandt's Bat.

The fur is long, somewhat coarse, and dark gray at the base. The coloration of the back varies greatly, from dark nut brown or dark gray brown to, rarely, light brown. As a rule, it is always darker than Brandt's Bat. The underside is a dark to light gray.

Juveniles are darker. The fur is black at the base and the back is a dark gray-brown. The wing membranes are relatively narrow, with the arm membrane attaching at the base of the toe. The feet are small. The spur is shorter than half the length of the tail membrane, and there is usually a narrow edging of skin present on the spur. The penis is thin, not thickened at its tip. The cusp next to the third premolar is lower than the cusp beside the second premolar. The second premolar is clearly smaller than the first premolar.

This is the liveliest species of *Myotis* in temperament.

Anomalies in coloration: Albinism.

Similar species: Brandt's Bat. (please refer to that species)

The Common Pipistrelle has a different ear and tragus shape, and the spur has an epiblema.

Range: this species is found throughout Europe, except for Scotland and northern Scandinavia. In the north it reaches approximately 65° latitude. The center of distribution is in central Europe. In the south, it occurs in the Balkan and Mediterranean region, but has not been documented from Spain.

Biotope: This species is not as closely tied to forests and bodies of water as Brandt's Bat, but is found more often in parks, gardens, and towns. It is more of a house than a forest bat. In southeastern Europe it is also found in karst formations. In summer, it occurs up to 1923

Whiskered Bat.

meters in altitude (Col de Bretolet, Alps), in winter up to 1800 meters (Tatra).

Summer roosts (maternity roosts) are usually located in buildings in narrow cracks accessible from the outside, between joists and foundations, behind shutters, and, rarely, in nest boxes.

Winter roosts are located in caves, tunnels, and cellars, with temperatures of 2 to 8°C. Usually it roosts free hanging on the wall or ceiling, but it is also found wedged into cracks. Winter

Whiskered Bat (left) and Brandt's Bat (right). The differences in coloration between the adults of both species are not always this pronounced.

roosts with more than 100 individuals are very rare. Males often predominate in the roosts.

It hibernates from October to March.

Migrations: It is probably mainly a permanent resident, but possibly can be classified as a partial migrant. The longest migration noted is 240 kilometers.

Reproduction: Females reach sexual maturity in the first year.

Mating takes place from fall through the time in the winter roost and into the spring.

Maternity roosts are occupied by 20 to 70 females starting in about May; during this time the males are solitary.

Females give birth to one

youngster in about mid June.
Maternity roosts break up in late
August.

Maximum age: 19 years;
average age 3.5 years.

Hunting and diet: This bat
emerges from its roost in early
evening.

It hunts at altitudes of 1.5 to 6
meters in parks, gardens, and
over-flowing bodies of water, but
it also hunts over meadows or in
forests. It hangs from branches
during pauses in hunting. Its
flight is fast, agile, and twisting.
In spring and fall it occasionally
hunts during the day.

Prey include mosquitoes,
mayflies, small dragonflies, small
beetles, and moths.

Calls: When disturbed, it emits
a persistent high-pitched
scolding or twittering.

The echolocation call
(searching flight) is an FM call
from 75 to 32 kHz (see
sonogram). Its duration is 2.5 to
3 ms, with a maximum impulse
intensity at 40 to 50 kHz. Calls
are repeated at 90 to 100 ms
intervals (about 10 to 11 calls per
second). The range is 5 to 20 m.

Note: In Bulgaria, Croatia,
Bosnia, Serbia, Montenegro, and
Greece, a very similar species (or
subspecies), *Myotis mystacinus
przewalskii* (Bobrinski, 1926),
can be expected. Its physical
measurements are somewhat
smaller, there is no cusp next to
the third premolar, as in *M.
mystacinus* and *M. brandti,* and
the second premolar is clearly
smaller than the first premolar,
and displaced to the inside.

Protection: This species is

severely endangered in western
Germany, and endangered in
eastern Germany and Austria.
there should be systematic
protection of the known
maternity and winter roosts.

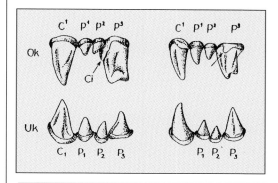

Brandt's Bat (left) and Whiskered Bat
(right). Shown here are teeth that are
important for species identification in
the upper mandible (OK) and lower mandible (UK). The rows of teeth are viewed
from the inside. c: canine; P: premolar.
Note the length of the secondary cusp
Ci (cingulum cusp) on the third premolar
in comparison to the second premolar.

MYOTIS EMARGINATUS (GEOFFROY, 1806)

Eng.: Geoffroy's bat
Ger.: *Wimperfledermaus*
Fr.: *Vespertilion à oreilles
échancrées*
Head-body: 41-53 mm
Tail: 38-46 (48) mm
Forearm: 36-41 (42) mm
Ear: 14-17 mm
Wingspan: 220-245 mm
Condylobasal length: 14-15.7
mm
Weight: (6) 7-15 g
Distinguishing characters:
Medium sized. The ears are of
medium length, with the upper
third of the outside margin

having a distinct, nearly right-angled indentation, and six or seven horizontal creases. the tragus is lanceolate, with more or less distinct, fine notches on the outside margin that do not quite reach the height of the indentation on the outside margin of the ear.

Juveniles are considerably darker, smoky gray to gray-brown in color, and without reddish tints. The muzzle is rufous and the ears and wing membranes are dark gray-brown. The wings are relatively broad, with the arm membrane attaching at the base of the toe. The feet are small. The spur is straight, attaining about half the length of the tail membrane; the free margin of the tail membrane has sparse, short, straight, soft hairs ("lashes").

Anomalies in coloration: Unknown.

Similar species: Natterer's Bat is grayer in color, its spur is S-shaped, and the free margin of the tail membrane is densely covered with stiff, curved bristles.

Brandt's Bat has the tragus projecting beyond the indentation on the hind margin of the ear, and the hairs on the back are not three colored.

For Daubenton's and Bechstein's Bats, see the diagnostic key.

Range: Occurs in central and southern Europe, including northeastern Spain, France, Belgium, southern Holland, rarely in the south of western Germany (Bavaria, absent in northern western Germany and

eastern Germany), but more abundant locally in the Czech Republic, Slovakia, southern Poland, the Ukraine, Italy, and the Balkans.

Biotope: This is a warmth-loving species. In the north it is predominately a house bat, in the south it is a cave bat. It occurs in lowlands and at lower elevations in mountains, both in towns with parks, gardens, and water and in karst formations, and at altitudes up to 1000 meters in mountains.

Summer roosts (maternity roosts) are located in warm attics (36 to 40°C).

It hangs in the open from rafters or the ridge of a roof. In southern Europe it is found mainly in caves and tunnels.

Winter roosts are in caves, tunnels, and cellars, with the temperature 6 to 9°C, rarely lower. Usually it is free hanging singly on the ceiling or wall, rarely in small clusters or in cracks.

It hibernates from October to March or April.

Migrations: This species is primarily a permanent resident with migrations usually less than 40 kilometers. The longest migration noted was 106 kilometers.

Reproduction: Females are already capable of breeding in the first year, but no births have yet been documented in the first year.

The mating season begins in the fall; whether or not mating occurs in the winter roost is unknown.

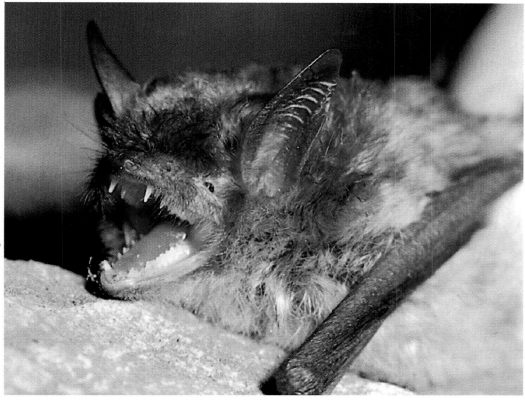

Geoffroy's Bat.

Maternity roosts are formed about May, often in the same roost as horseshoe bats. Maternity roost size varies from 20 to 200 females (Czech Republic, Slovakia) to 500 to 1000 females (France, Balkans).

Females give birth from mid or late June to early July to one youngster, which is capable of flight at about four weeks. Maternity roosts disband in September.

Maximum age: 16 years. The average age is 2.8 to 3.5 years.

Hunting and diet: It emerges from the roost in the early evening.

It hunts at altitudes of 1 to 5 meters, also over water. The flights are twisting. It preys primarily on spiders, but also dipterids, such as mosquitoes, and butterflies and caterpillars. Apparently the prey is also taken from branches as well as from the ground.

Calls: It emits a loud, shrill scolding in the maternity roost.

Protection: Geoffroy's Bat is threatened with extirpation in western Germany and endangered in Austria, in part with severe population decline. Systematic protection of roosts is necessary and caution must be exercised with wood preservatives!

MYOTIS NATTERERI **(KUHL, 1818)**
    Eng.: Natterer's Bat
    Ger.: Fransenfledermaus

Geoffroy's Bat.

Fr.: Vespertilion de Natterer
Head-body: (40) 42-50 (55) mm
Tail: 38-47 (49) mm
Forearm: 36.5-43.3 (46) mm
Ear: 14-18.3 (20) mm
Wingspan: 245-280 mm
Condylobasal length: 14-15.6 mm
Weight: 5-12 g
Distinguishing characters:

Medium sized. The ears are relatively long. The outside margin has five horizontal creases and a distinct indentation, which the long, lanceolate tragus projects beyond. The tragus is longer than half the ear length. The muzzle is relatively long, and the upper lip has a suggestion of a mustache of fairly long hairs. The fur is long, sparse, and dark gray at the base. The back is light gray, with only a slight brownish tinge. The underside is light whitish gray, sharply set off from the back. The muzzle is light flesh colored, the ears and wing membranes light gray brown, and the tragus is light yellowish gray, becoming darker toward the tip. The wing membrane is broad, and attaches at the base of the toe. The S-shaped spur attains about half the length of the tail membrane; the free margin of the tail membrane is wrinkled and densely covered with two rows of stiff, downward curving bristles ("fringes"). The feet are small.

Anomalies in coloration: Unknown.

Similar species: See the species descriptions and diagnostic key for comparisons with the Long-fingered Bat, Daubenton's Bat, the Pond Bat,

Natterer's Bat.

Geoffroy's Bat, and Bechstein's Bat. However, none of these species has an S-shaped spur.

Range: Occurs throughout almost all of Europe, in northern Ireland, England, Denmark, southern Sweden, Estonia, and in the Mediterranean region, but has not been documented from the Balkans.

Biotope: This is primarily a forest bat. It occurs in forests and parks with bodies of water and wetlands, as well as in towns. It has been documented in the summer up to a 1923 meter altitude (Col de Bretolet, Alps), in winter only up to 800 meters.

Summer roosts (maternity roosts) are located both in tree holes and bat boxes in woodlands, as well as in cracks on or in buildings (on rafters, more rarely on the roof ridge). Individual animals also roost in cracks under bridges and behind shutters.

Winter roosts are located in tunnels, caves, and cellars; the temperatures are 2.5 to 8°C and the humidity is high. They are usually found wedged into tight cracks, in part also lying on the back; also occur in floor rubble or free hanging on the ceiling or the wall. Occasionally they roost in small clusters, then usually intermingled with Daubenton's Bat.

It hibernates from October to April. Usually it enters the winter roost later than Daubenton's Bat.

Migrations: Natterer's Bat is a permanent resident. Its longest migration is 90 kilometers.

Reproduction: Accurate information about the onset of sexual maturity is lacking.

Mating possibly takes place from fall to spring (?).

Maternity roosts are occupied from April to May by 20 to 80 females. Individual males may also be present in the maternity roosts.

Females give birth to one youngster from mid June to early July. Maternity roosts are sometimes changed very frequently (as often as once or twice a week).

Maximum age: 17 years, 5 months.

Hunting and diet: This species emerges from the roost in late evening. Its flight is low (1 to 4 meters), slow, and in part with whirring wing beats. It is very maneuverable in confined spaces and can also hover briefly. It hunts in woodlands as well as over water. It usually hunts the entire night.

Natterer's Bat primarily preys on diurnal flies and other dipterids, which apparently are harvested from leaves, branches, and other substrates during their nocturnal resting phase.

Calls: Usually a chirping or peeping, lower in pitch than Daubenton's Bat. When disturbed, it also emits a low-pitched humming sound. High-pitched, shrill calls are given in flight.

Echolocation calls (searching flight) are FM calls from 78 to 35 kHz. The duration is 2 ms (see the sonogram). The maximum impulse intensity is at 50 kHz,

Hibernating Natterer's Bats. Readily visible are the relatively long ears that project beyond the tip of the muzzle.

and the calls are repeated at 70 to 90 ms intervals (11 to 14 calls per second). The range is 5 to 20 meters. The calls are softer than those of *M. mystacinus*.

Protection: This species is severely endangered in western Germany; in eastern Germany, as far as can be determined, the populations are relatively stable. It is threatened by the loss of its roosts (felling of hollow trees, destruction of the winter roosts, etc.).

MYOTIS BECHSTEINI (KUHL, 1818)
Eng.: Bechstein's Bat

Ger.: *Bechsteinfledermaus*
Fr.: *Vespertilion de Bechstein*
Head-body: 45-55 mm
Tail: (34) 41-45 (47) mm
Forearm: 39-47 mm
Ear: (21) 23-26 mm
Wingspan: 250-286 mm
Condylobasal length: 16-16.8 mm
Weight: 7-12 g

Distinguishing characters: Medium sized. The ear is conspicuously long and rather broad. When tilted forward, they project beyond the muzzle. The outside margin of the ear has 9 horizontal creases. The tragus is long and lanceolate, reaching to about half the ear's length. The fur is relatively long and dark gray-brown at the base. The back is pale brown to reddish brown. The underside is light gray. Juveniles are light gray to ashy gray. The muzzle is rufous, the ears and wing membranes are light gray brown. The wings are broad and short, with the arm membrane attaching at the base of the toe. The feet are small. the spur is straight, reaching about one-third to one-half the length of the tail membrane. The last tail vertebra is free.

Anomalies in coloration: Partial albinism (there are white wing tips on both sides).

Similar species: the Long-eared Bats have longer ears that touch in front at the base.

All *Myotis* species of approximately the same size have considerably shorter ears.

Range: Occurs in the temperate zones of Europe. Apparently it is only locally distributed and not abundant anywhere. It is found in southern England, France, Belgium, the Netherlands, Germany (does not reach the Baltic Sea coast), southern Sweden (?), the Mediterranean and Balkan regions, and in Bulgaria on the Black Sea coast.

Biotope: This is a forest bat. It is primarily found in moist mixed forests, but also occurs in coniferous forest, parks, and gardens in lowlands and low mountains. In the summer it occurs to an altitude of 800 meters, in the winter to 1160 meters. Summer roosts (maternity roosts) are located in tree holes and bat boxes (not shallow boxes), rarely in buildings, where it roosts free hanging. Individual animals also roost in rock cavities.

Winter roosts are found in cellars, tunnels, and caves, and possibly also sporadically in tree holes. The temperatures range from 3 to 7°C, and the humidity is high. It is often free hanging on the ceiling or wall, more rarely found wedged into tight cracks. It usually hangs singly, not in clusters. The ears point straight out even in hibernation.

It hibernates from October to March or April.

Migrations: Apparently this species is a permanent resident. Its longest migration is 35 kilometers.

Reproduction: The time of the onset of sexual maturity is unknown.

The mating season extends from fall to spring (?). Maternity roosts, which are changed frequently, are occupied beginning in late April to May by 10 to 30 females.

Females give birth to a single youngster from late June to early July. Juveniles are capable of flight in early to mid August. Maternity roosts break up in late August. Males are solitary in summer.

Maximum age: 21 years.

Hunting and diet: This bat emerges from the roost only in total darkness. It has a fluttering flight, but is very agile even in the most confined space. It hunts low (1 to 5 meters), also taking prey from twigs or the ground (?). It preys on moths, mosquitoes, and beetles.

Bechstein's Bat.

Calls: When threatened, it emits a dull humming or chirping sound. No audible calls are produced in flight.

Two kinds of echolocation calls (searching calls) can be differentiated:

I. A steeply falling, short FM call from 80 to 38 kHz with a duration of 2 to 2.5 ms (see the sonogram). Calls are repeated at 65 to 100 ms intervals (about 10 to 15 calls per second).

II. A flat, fairly long FM call from 60 to 32 kHz with a duration of 4 to 5 ms (see sonogram). Calls are repeated at 100 to 110 ms intervals (about 9 to 10 calls per second).

Protection: This species is endangered in western Germany and Austria, and severely endangered in eastern Germany. Because it is not common anywhere, systematic protection of known maternity roosts and installation of bat boxes are essential.

MYOTIS MYOTIS (BORKHAUSEN, 1797)

Eng.: Greater Mouse-eared Bat
Ger.: *Grosses Mausohr*
Fr.: *Grand Murin*
Head-body: (65) 67-79 (84) mm
Tail: (40) 45-60 mm
Forearm: 54-67 (68) mm
Ear: (13) 26-31 mm
Wingspan: 350-430 mm

Partially albino Bechstein's Bat with white wing tips.

Condylobasal length: (21.5) 22-24 mm
Weight: (20) 28-40 g
Distinguishing characters: Large species. The muzzle is short and wide. The ears are long and broad, the outside margin with 7 to 8 horizontal creases. The front margin of the ear clearly curves backward; the tip of the ear is broad. The tragus is broad at the base, reaching nearly half the ear's length.

The fur is dense and short, black-brown at the base. The back is a light gray-brown, in part with a rusty tinge; the underside is whitish gray. The muzzle, ears, and wing membranes are brown-gray. The juveniles are darker, smoky gray without a brownish tint. The

Greater Mouse-eared Bat. When bats feel threatened, they open their mouths wide and display their sharp teeth. They also often emit defensive calls at the same time.

Greater Mouse-eared Bat.

wings are broad, with the arm membrane attaching at the base of the toe. The spur reaches about half the length of the tail membrane. There is a narrow edging of skin.

Anomalies in coloration: Partial albinism (white wing tips).

Similar species: This species can only be confused with the Lesser Mouse-eared Bat, but in this species the ear is distinctly narrower and shorter (less than 26 millimeters long), the front margin of the ear is straighter, the ears are more pointed, and the tragus is narrower. The muzzle is narrower and more pointed, appearing longer. The whole animal has a more graceful appearance than the Lesser Mouse-eared Bat.

Range: The Greater Mouse-eared Bat occurs in central and southern Europe, but is absent from Ireland, Denmark, and Scandinavia; it is practically extirpated in England; occurs on the northern border of France, in Belgium, the Netherlands (threatened with extirpation), western Germany (northernmost locality Husum, Schleswig-Holstein), eastern Germany (northernmost winter roost at Sassnitz on the Island of Rügen), also reaches Poland on the Baltic Sea coast. In southern Europe it is found in the Balkans and the Mediterranean region.

Biotope: This is a warmth-loving species. It is a house bat in the north, a cave bat in the south. It is found in the open countryside, landscapes with open stands of trees, and parks; in the north it is tied to human settlements. It is usually found at altitudes below 600 meters, but in the summer up to 1923 meters (Col de Bretolet, Switzerland), in the winter up to 1460 meters (Tatra).

Summer roosts (maternity roosts) in the north are located in warm attics, church steeples (temperature up to 45°C), rarely in warm underground spaces;

while in the south it is found in caves. Individual animals also roost in nest boxes or tree holes.

Winter roosts are located in caves, tunnels, and cellars, with temperatures (3°) 7° to 12°C. It almost always roosts free hanging, but often protected in ceiling shafts, holes in the wall, or hollow spaces in ceilings, rarely in narrow cracks. It often forms clusters (up to more than 100 individuals). Formerly, certain winter roosts contained up to 4500 individuals, but today they rarely have more than several hundred animals.

It hibernates from September or October to early March or April. Females appear in the roosts before the males. A change in roosts is possible in the winter. The hibernation phases last up to six weeks.

Migrations: This species is a partial migrant. The distance between summer and winter roosts in the north is about 50 kilometers. Migrations of more than 100 kilometers are not uncommon. The longest migration was 390 kilometers (Spain).

Reproduction: A small percentage of females are ready to breed in the first year.

Breeding starts in August. One male can have a harem of up to five females. Mating in the winter roost is also possible. Maternity roosts are found up to an altitude of 1016 meters. Northernmost maternity roosts in eastern Germany are located about 60 kilometers from the Baltic Sea. The maternity roosts are occupied beginning in March by a maximum of 2000 females. At this time the males are solitary; individual males rarely occur in maternity roosts.

Females give birth starting in early June, usually in the morning hours. The mothers fly out alone to hunt the same evening, the young being left behind in groups. Often a few females stay behind with them. Newborns are pink, with fine, barely pigmented hairs on the back. *First day:* weight approximately 6 grams, forearm length 22.6 millimeters. *Eleventh day:* weight 13 grams, forearm length 41.9 millimeters. *Twenty-second day:* weight 18-19 grams, forearm length 55.4 millimeters. The eyes open on the fourth to seventh day, and the coat is complete on the 22nd day. The permanent teeth are fully developed on the 13th to 35th day. They are capable of flight at 20 to 24 days of age, and independent at 40 days (mid July). There is a high infant mortality in cold weather (40% or more).

Maximum age: 22 years. The average age is 4 to 5 years.

Hunting and Diet: This bat usually does not emerge from its roost until after dark.

It hunts in parks, in fields and meadows, as well as in towns. The flight is slow with long wing beats, at 5 to 10 meters in altitude, but in part just above the ground. It often hunts "on foot" on the ground. It preys predominantly on ground beetles, along with June beetles, carrion beetles, grasshoppers, crickets, moths, and spiders.

Calls: In the maternity roost it emits a loud, shrill scolding and screeching call; when threatened it also emits a low-pitched humming (like large bumblebees), and when disturbed in hibernation there is a persistent, loud screeching.

Echolocation calls (searching flight) are FM calls from 62 to 28 kHz (105 to 30 kHz), with a duration of 2 to 3 ms. The calls are repeated at 50 to 90 ms interval (approximately 12 to 20 calls per second) (see sonogram).

Protection: This species is severely endangered in western Germany and Austria, and in eastern Germany it is listed among the animal species threatened with extirpation. In recent years, however, there has been a stabilization of the remnant populations in many regions. Almost everywhere in central Europe there have been population declines of 80% or more in the last 20 to 30 years.

Systematic protection of summer and winter roosts, avoidance of wood preservatives that are toxic to warm-blooded animals, and protection of the biotope are necessary. Leave entrance openings (30 x 30 centimeters) in roosts; if necessary also provide enough slits.

MYOTIS BLYTHI (TOMES, 1857)
Eng.: Lesser Mouse-eared Bat
Ger.: *Kleines Mausohr*
Fr.: *Petit Murin*
Head-body: (54) 62-71 (76) mm

Tail: 53-59 (60) mm
Forearm: 52.5-59 (61.5) mm
Ear: 19.8-23.5 (26) mm
Wingspan: 380-400 mm
Condylobasal length: 17.2-18.5 mm
Weight: 15-28 g

Distinguishing characters: This species is very similar to the Greater Mouse-eared Bat, except that it is somewhat smaller. The ears are narrow and somewhat shorter than in the Greater Mouse-eared Bat, the front margin of the ear curves rearward less strongly, the ear tapering to more of a point. The tragus is narrower at the base, lanceolate, and reaches almost half the ear length. The outside margin of the ear has five or six horizontal creases. The muzzle is narrower and more pointed than in the Greater Mouse-eared Bat, making it appear longer.

The fur is short, dark gray at the base. The back is gray with a brownish tinge. the underside is grayish white. The muzzle, wing membranes, and ears are light gray-brown, the tragus pale yellowish white. The arm membrane attaches at the base of the toe. The spur reaches about half the length of the tail membrane and has a narrow edging of skin.

Anomalies in coloration: Unknown.

Range: This species occurs in southern Europe and the Mediterranean region. the northern limit of its range: Spain, southeastern France, sporadically in Switzerland and Austria, Slovakia, Hungary, Romania, Moldavia, and the Ukraine.

Protection: The Lesser Mouse-eared Bat is threatened with extirpation in Austria. There also are reports of population declines in southeastern Europe.

Protection of the caves and other roosts is necessary.

Biotope: This bat prefers warm regions with open stands of trees and shrubs, karst formations, parks, as well as towns. Any ecological differences in comparison to the Greater Mouse-eared Bat are not precisely known; both species occur in the same range. It is documented up to an altitude of 1000 meters.

Summer roosts (maternity roosts) are located primarily in warm caves, often together with Schreiber's Bat and horseshoe bats. It also occupies warm attics, free hanging on the roof ridge. Individual animals are rarely found in tree holes.

Winter roosts are in caves and tunnels. The temperatures range from 6° to 12°C. Roosts are primarily free hanging.

Migrations: Apparently this species is a partial migrant. Its longest migration was 600 kilometers (Spain).

Reproduction: Mating occurs in the fall, apparently also into the spring. A male can have a harem of females. Large maternity roosts of up to 5000 individuals are formed. In attics, mixed maternity roosts with the Greater Mouse-eared Bat are known. Females give birth to a single youngster.

Lesser Mouse-eared Bat.

Lesser Mouse-eared Bat.

Maximum age: 13 years.

Hunting and diet: This bat emerges from its roost in late evening or after dark. Its flight is slow, regular, and more agile in a confined space than the Greater Mouse-eared Bat. Apparently it also takes prey from the ground. It preys on moths and beetles.

Calls: The calls are similar to those of the Greater Mouse-eared Bat. When defending itself, it emits a loud, shrill scolding or low-pitched humming.

Genus: *Nyctalus* (Bodwich, 1825)

There are six species in the genus *Nyctalus*, three of which occur in Europe. They are medium-sized to large, robust forest bats with brown to rufous fur. Their ears are short and triangular, the tragus is broad, like a mushroom. The spurs have the epiblema divided by a clearly visible steg. The wings are long and pointed. There are two teats.

$$\text{tooth formula} \quad \frac{2123}{3123} = 34$$

NYCTALUS NOCTULA (SCHREBER, 1774)

Eng.: Noctule
Ger.: *Grosser Abendsegler*
Fr.: *Noctule*
Head-body: 60-82 (84.8) mm
Tail: 41-60.6 mm
Forearm: 48-58 mm
Ear: 16-21.1 mm
Wingspan: 320-400 mm
Condylobasal length: 17.4-19.9 mm
Weight: (17) 19-40 (46) g

Distinguishing characters: Large. The ears are broad, triangular, rounded at the tip, the outside margin with four to five horizontal creases, and very wide at the base. The tragus is short, shaped like a mushroom. Glandular bumps are visible at the corners of the open mouth.

The fur is short, close lying, of a single color. The back has a rufous sheen in the summer. The underside is a dull, lighter brown. After the molt (August and September), the back is a dull, pale brown, in part with a light gray tinge. Juveniles are dull brown on the back, altogether darker. The ears, muzzle, and wing membranes are blackish brown. the wings are long and narrow, with the arm membrane attaching at the ankle. The spur reaches half the length of the tail membrane. The broad epiblema has a visible steg present. When excited, this bat emits a strong musky odor. When disturbed during its daytime lethargy, it may go into akinesis, as does the Common Pipistrelle.

Anomalies in coloration: Albinism.

Similar species: there is a clear difference in size in comparison to Leisler's Bat and the Greater Noctule. Leisler's Bat also has two-toned fur on its back.

The Serotine is darker, its tragus not mushroom shaped, the epiblema narrow, and the last one or two vertebrae are free.

Range: The Noctule occurs throughout Europe, except for Ireland, Scotland, and northern Scandinavia. In the north it reaches approximately 60° latitude; in the south it is also present in the Balkan and Mediterranean regions.

Biotope: This is a forest bat, but it also occurs in larger parks. It lives primarily in lowlands, but during migrations also up to an altitude of 1923 meters (Col de Bretolet, Alps).

Summer roosts (maternity roosts) are located in tree holes (woodpecker holes or holes produced by rot, cracks in trunks). The entrance holes are round or slit-shaped, 6 centimeters in diameter; the cavity diameter is about 12 centimeters, and the height of the entrance hole above the ground

The Noctule (left) is one of the largest European bat species. On the other hand, the Whiskered Bat (right) is one of the smallest species.

varies from 1 to 20 meters. It also occupies bat boxes. In summer it has also been observed in hollow concrete light poles and in cracks between the concrete panels of apartment houses.

Winter roosts are located in thick-walled tree holes, deep rock crevices, and cracks in the walls of houses; in southeastern Europe also in caves. In cities it has been found in ventilation shafts of apartment buildings and in churches. Up to 1000 individuals can be found in large winter roosts. The temperature is usually low, even down to approximately 0°C for short periods.

Hibernation occurs from early October or mid November to mid March or early April. Outside of winter roosts in cracks in rock or walls loud, shrill calls can frequently be heard, even at temperatures below 0°C. In mild weather, individual animals emerge. Noctules form clusters in the roosts, sometimes overlapping one another like roof tiles. In hard winters up to 50% of the individuals can freeze in unfavorable roosts.

Migrations: This is a migratory species. In central Europe the fall migration occurs from early September to mid November. The principal migration direction is to the southwest. The longest migration is 930 kilometers (in Russia, however, 1600 kilometers). Sometimes it also migrates during the day and has been observed together with swallows. In eastern Germany overwintering Noctules apparently migrate from northeastern Europe. Whether portions of the native populations are sedentary is questionable.

Reproduction: The Noctule reaches sexual maturity in the first year, but some females do not bear young until the second year. Males breed in the second year.

The mating season is from August to October. A male occupies a breeding roost (usually in tree holes) for several weeks and drives off other, sexually mature males. He emits mating calls at the entrance or in flight. He has a harem of 4 to 5 (20) females, which stay with the male for one or two days. Males that do not participate in breeding live in groups. Starting in April, males and females occupy summer roosts. Females move to maternity roosts starting in mid May. Each maternity roost is occupied by 20 to 50 (up to 100) females. Males stay in small groups outside the maternity roosts.

Females give birth from mid June to early July. In central Europe, litters, as a rule, consist of two, rarely three, young (in England usually one). The number of twin births is lowest in the southwest and increases toward the northeast. *Newborns:* naked, pink in color, weight 7.5 grams, forearm length 20 millimeters. *Eleventh day:* weight 10 grams, forearm length 32 millimeters. The eyes open between the third and sixth day, and the young have a complete coat of gray to silvery gray hair. At 36 days the color changes to brown. Juveniles are capable of flight starting in the fourth week. Permanent teeth are complete starting in the fifth to seventh week, at which time they also become independent.

Maximum age: 12 years.

Hunting and diet: This bat emerges from the roost early, sometimes before sunset. The hunting flight lasts 1 to 1.5 hours. In the summer there is often a second hunting flight before sunrise. The flight is fast (up to 50 kilometers per hour), high, at altitudes of 10 to 40 meters (up to 70 meters), and straight, with quick turns and dives. In flight the wings nearly touch under the body. The flight silhouette displays long, slender, often distinctly angled wings and a keel-shaped tail membrane, which has a smooth rear margin when spread.

It hunts over meadows, lakes, garbage dumps, as well as above the treetops. The hunting territory is located up to 6 kilometers from the roost. It preys on moths, June beetles, and other large flying insects.

Protection: The Noctule is endangered in western Germany and Austria; it is endangered in eastern Germany, but not uncommon in suitable biotopes.

It is threatened through the cutting down of hollow trees (maternity roosts and winter roosts) and the destruction of winter roosts in buildings. If a suitable alternate roost is not immediately available following the destruction of the winter roost, the animals must be

Noctule. As with all common bats, the crack in the eyelid is barely visible with closed eyes.

overwintered artificially in suitable quarters (for example, in a zoo). The lack of hollow trees for maternity roosts can be offset through the use of bat boxes.

figure, page 137

Calls: In maternity roosts or in winter roosts, there is a high-pitched loud scolding or twittering at intervals.

Two types of echolocation calls (searching flight) can be distinguished (see sonogram)

1. An FM call from 45 to 25 kHz, with a duration of 6 ms; the maximum impulse intensity is at 25 kHz. Calls are repeated at 125 ms intervals (= 8 calls per second). These calls are used in hunting at low altitude.

2. A long, flat call from 25 to 19 kHz, with a duration of 25 ms. The calls are repeated at 300 to 400 ms intervals (= 3 calls per second). These calls are used in hunting above tree level.

Both types of calls are usually alternated regularly. They are very loud calls and range up to 150 meters.

Social calls: In flight these are loud, short, metallic-sounding calls ("tzick" or "bick") audible up to 50 meters away, which are almost painful to the ear in the immediate vicinity. They are 32 to 17 kHz; rising and falling 14 times in 110 ms.

NYCTALUS LEISLERI (KUHL, 1818)

Eng.: Leisler's Bat
Ger.: *Kleiner Abendsegler*
Fr.: *Noctule de Leisler*
Head-body: 48-68 mm
Tail: 35-45 mm
Forearm: (37) 39-46.4 mm
Ear: 12-16 mm
Wingspan: 260-320 mm
Condylobasal length: 14.7-16 mm
Weight: 13-20 g

Distinguishing characters: Medium sized. The external ear and tragus are as in the Noctule, the outside margin of ear having four or five horizontal creases. The muzzle is somewhat more pointed.

The fur is short, the hairs bicolored, blackish brown at the base. The back is rufous, usually somewhat darker and less shiny than in the Noctule. The underside is yellow-brown. The face, ears, and wing membranes are blackish brown. Juveniles are altogether darker. Wings are long and narrow, the wing membranes with dense hair along the body and the arms. The arm membrane attaches at the ankle. The spurs and epiblema are as in the Noctule.

Anomalies in coloration: Unknown.

Similar species: The Noctule is larger and its hairs are monotone.

Nathusius' Pipistrelle is smaller and its tragus is not mushroom-shaped.

Range: Leisler's Bat is found throughout almost all of Europe, but its occurrence is sporadic everywhere. It occurs in Ireland, southern England, France, the Netherlands, eastern and western Germany, Poland, the Ukraine, Belarus, and Russia. In the north it apparently does not reach the North Sea or Baltic coasts and is absent from Scandinavia. In the south it occurs in Portugal, Spain, Italy, and the Balkan region.

Biotope: This is a forest bat. As in the Noctule, in summer it is found up to an altitude of 1923 meters (Col de Bretolet, Alps).

Summer roosts (maternity roosts) are located in tree holes and bat boxes, in part together with the Noctule. More rarely it is found in cracks in buildings.

Winter roosts are located in tree holes, as well as in cracks and hollow spaces on and in buildings. It overwinters in large groups.

It hibernates from late September to early April.

Migrations: This is a migratory species. Its longest migration is 810 kilometers. The migration direction is apparently from the east to the southwest.

Reproduction: The time of the

Leisler's Bat.

onset of sexual maturity is unknown. The mating season is in late August and September. The male has a breeding roost and a harem of up to nine females. Maternity roosts of 20 to 50 females are located in tree holes. Up to 500 females can be present in buildings (Ireland).

Females give birth to two young starting in mid June.

Maximum age: 9 years.

Hunting and diet: Similar to the Noctule. It emerges from the roost shortly after sunset. Its flight is fast, high, and includes dives. It preys on moths, beetles, and other flying insects.

Calls: Its short, shrill calls resemble those of the Noctule.

Protection: This species is severely endangered in western Germany and is a rare species in eastern Germany; it is endangered in Austria.

Protective measures should be

Head-body: 84-104 mm
Tail: 55-65 mm
Forearm: 63-69 mm
Ear: 21-26 mm
Wingspan: 410-460 mm
First phalanx, fourth finger approximately 8.2 mm
Second phalanx, fourth finger:

Leisler's Bat.

the same as for the Noctule. Putting out bat boxes is important.

NYCTALUS LASIOPTERUS
(SCHREIBER, 1780)
  Eng.: Greater Noctule
  Ger.: *Riesenabendsegler*
  Fr.: *Noctule géante*

approximately 14-15 mm
  Condylobasal length: 22-23.6 mm
  Weight: 41-76 g
  Distinguishing characters: This is the largest European bat. Its ear similar to that of the Noctule, but wider; the back of the broadened ear margin has sparse hairs. The tragus is mushroom-shaped.

The fur is dense and relatively long, the hairs monotone. The back is rufous, usually darker than in the Noctule; the underside is a yellow brown. Juveniles are darker in color. The muzzle and ears are blackish brown, the wing membranes dark brown. The wings are long and narrow, with rusty hairs on the underside along the body. The attachment point of the arm membranes, spur, and epiblema are as in the Noctule.

Anomalies in coloration: Unknown.

Similar species: The Noctule, though it is distinctly smaller.

Range: Information on the range is sketchy. It apparently occurs primarily in southeastern Europe. It has been documented from Spain, Portugal, southern France, Italy, Switzerland, Austria, Slovakia, Poland, Hungary, Romania, Bulgaria, the Ukraine, the European part of Russia, Bulgaria, Croatia, Bosnia, Montenegro, Albania, and Greece. Usually there are only single reports.

Biotope: The Greater Noctule is a forest bat. It is common in deciduous forests. It has been documented at an altitude of 1923 meters (Col de Bretolet, Alps).

It forms summer roosts in Russia, in part together with the Noctule, Nathusius's Pipistrelle, and the Common Pipistrelle.

Maternity roosts and winter roosts are located in tree holes.

Migrations: This is a migratory species. Animals living in Belarus migrate to the southeast in the fall.

Reproduction: The time of the onset of sexual maturity and the breeding season are not precisely known, though they are probably similar to those of the Noctule. Only small maternity roosts (up to 10 females) are formed, in part together with the aforementioned species.

Females give birth starting in late June to two young, more rarely one. *Newborns:* weight 5 to 7 grams, forearm length 20 to 26 millimeters. *Tenth day:* forearm length 30 to 35 millimeters. *Twentieth day:* forearm length 40 to 45 millimeters. The forearm stops growing starting on the fortieth day. The hair length at this time is as in the adults. The young are also capable of flight at this time.

Maximum age: Unknown.

Hunting and diet: Hunting style and diet are not precisely known. They are probably similar to that of the Noctule.

Protection: As far as can be estimated, given the lack of information, the situation is the same as in the two other *Nyctalus* species.

EPTESICUS RAFINESQUE, 1820

The genus *Eptesicus* includes thirty species, two of which occur in Europe. The ears and muzzle are black. One or two tail vertebrae project beyond the tail membrane. Spurs have a narrow epiblema. Wings are relatively broad. Two teats are present.

tooth formula $\dfrac{2113}{3123} = 32$

Greater Noctule.

EPTESICUS SEROTINUS (SCHREBER, 1774)

    Eng.: Serotine
    Ger.: *Breitflügelfledermaus*
    Fr.: *Sérotine commune*
    Head-body: 62.6-82 mm
    Tail: (39) 46-54 (59) mm
    Forearm: 48-57 mm
    Ear: (12) 14-21.8 mm
    Wingspan: 315-381 mm
    Condylobasal length: 18-21.2 mm
    Weight: 14.4-33.5 (35) g

Distinguishing characters: Large species. The ears are relatively short, nearly triangular; the hind margin of the ear is narrow, and there are five horizontal creases running in the direction of the corner of the mouth, ending before it. The tragus attains about one-third of the ear length, curving slightly inward, and rounded on top.

    The fur is long, dark brown at the base. The back is a dark smoky brown, with some variation, and the tips of the hairs are in part slightly shiny. The underside is yellow-brown, the boundary between the back and the underside indistinct. Juveniles are altogether darker. The ears and muzzle are black, the wing membranes dark blackish brown. The wings are broad, the arm membrane attaching at the base of the toe. One or two tail vertebrae are free (4 to 5 millimeters). The spur attains about one-third to one-half the length of the tail membrane. A narrow epiblema without a visible steg is present. This species has a very calm temperament.

Anomalies in coloration: Albinism has been observed repeatedly.

Similar species: The Serotine can hardly be confused with any other species.

The Northern Bat and Parti-colored Bat are smaller, and the back is differently colored.

The Noctule is rufous, with a mushroom-shaped tragus.

Greater and Lesser Mouse-eared Bats have the back more gray, and the ear and tragus longer and differently shaped.

Range: The Serotine occurs throughout Europe. It ranges in the north to 55° latitude (southern England, Denmark, southern Sweden); in the south it occurs in the Mediterranean and Balkan regions to the Caucasus.

Biotope: This is a house bat. It is primarily found in lowlands, in areas with human settlement with parks, gardens, meadows, and on the outskirts of cities. It has been documented up to altitudes of 900 meters (in summer) and 1100 meters (in winter).

Summer roosts (maternity roosts) are frequently in the roof ridges of attics, usually not free hanging, but hidden under the roof laths or beams. Individual animals (usually males) also occur on joists, behind shutters, and rarely in bird or bat boxes. In southeastern Europe it is also found in limestone caves. Winter

Serotine.

The some-what equal-sized Broad-winged Bat (a, b) and larger Evening Bat (c, d) can be identified with some practice. The smaller Evening Bat exhibits a narrower, often distinctly angled wing and a wedge-shaped tail. The Broad-winged Bat has broad wings and a short, rounded or many pointed tail (*after Klawitter and Vierhaus, 1975*).

roosts are located in caves, tunnels, cellars, as well as deep in roof joists of attics, behind pictures in churches, and in wood piles. Although this is a rather abundant species, no mass roosts are known. Predominantly individual males are found, both wedged into cracks and free hanging on the ceiling or wall. They have also been found in the floor rubble. Temperatures range from 2 to 4°C, and the humidity is relatively low. It hibernates approximately from October to late March and April.

Migrations: This bat is difficult to classify, although it is probably more of a permanent resident. Documented migrations of 83, 145, 204, and 330 kilometers, however, also argue in favor of a partial migrant.

Reproduction: Females become sexually mature in the first year (in Russia).

The mating season begins in late August. Whether or not it continues into the spring is unknown. Maternity roosts are occupied beginning in April and May by 10 to 50 (100) females. When disturbed the bats run very quickly in the hollow spaces of the roof ridge. Males are solitary throughout the year.

Females give birth starting in the second half of June to one youngster in central Europe (in central Asia to two, rarely three young). Newborns: weight 5.2 to 6.2 grams, forearm length about 21 millimeters. The weight doubles in about 10 days. The forearm stops growing at about five weeks. The eyes open at seven to eight days. The permanent teeth are complete at the end of the third week, at which time the juveniles are also capable of flight. The juveniles are independent at about five weeks (late July to early August). The maternity roosts break up in late August.

Maximum age: 19 years, 3 months.

Hunting and diet: Serotines emerge from their roost in early evening. The flight is slow (15, maximum 30 kilometers per hour), at about 6 to 10 meters altitude in long arcs in gardens, along forest edges, over garbage dumps, and around street lamps. In flight it is usually silent. The behavior, it seems possible that they also take them from branches or the ground.

Serotines also emerge to hunt during a light rain.

Calls: When defending itself, it emits loud, high-pitched "chirping" or "tzicking" calls.

Echolocation calls (searching

Serotine.

broad wings are barely angled and the tail membrane appears rounded, not keel-shaped, short, and uneven. Occasionally there are two hunting flights a night. It rarely ranges more than one kilometer from its roost to its hunting territory.

It preys on moths and beetles. Because Serotines in captivity react to the imitated running sounds of beetles with searching

flight) are FM calls from (67) 52 to 25 kHz with a duration of 13.5 ms (see sonogram). The maximum impulse intensity is at 25 kHz. Calls are repeated at 150 ms intervals (= 6.7 calls per second). The range is 20 to 50 meters.

Protection: This species is severely endangered in western

Germany, endangered in eastern Germany and Austria.

It is particularly threatened by disturbances and the loss of maternity roosts. Avoid the use of wood preservatives that are toxic to mammals.

EPTESICUS NILSSONI (KEYSERLING ET BLASIUS, 1839)

Eng.: Northern Bat
Ger.: *Nordfledermaus*
Fr.: *Sérotine de Nilsson*
Head-body: (45) 54.5-63.5 mm
Tail: 35-50 mm
Forearm: (37) 38.1-42.8 (44) mm
Ear: (11.5) 13.8-17.3 mm
Wingspan: 240-280 mm
Condylobasal length: 14-15.2 (15.6) mm

Northern bat.

Weight: (6.5) 8-17.5 g
Distinguishing characters: Medium sized. The ears are relatively short, the hind ear margin with five horizontal creases, broadening toward the base, and reaching nearly to the corner of the mouth. The tragus is short, broad, curving slightly inward, and rounded on top.

The fur is long, dark brown at the base. On the back there are hairs with gleaming gold tips, as well as gleaming gold hairs on the crown of the head. The nape is darker; only here is there is a relatively sharp boundary with the yellow-brown underside. Juveniles are altogether darker, the back without a golden tinge, the tips of the hairs appearing more silvery, and the belly gray.

Northern bat.

The muzzle, ears, and wing membranes are blackish brown. The spur attains about half the length of the tail membrane. A narrow epiblema without a visible steg is present. The last tail vertebra is free (3 to 4 millimeters). The arm membrane attaches at the base of the toe.

Anomalies in coloration: Unknown.

Similar species: Savi's Pipistrelle: An overlap in size is scarcely possible. The tragus becomes wider above, the hind margin of the ear is scarcely widened, and the belly and throat are more white.

Parti-colored Bat: The hind margin of the ear is greatly widened, and has a different shape. The throat is white, the tips of the hairs on the back are silvery, and there is a distinct epiblema with a visible steg.

Serotine: This species is larger and differently colored.

Range: The Northern Bat is found in northern, central, and eastern Europe. It is the only European bat to reach the Arctic Circle. It occurs throughout Scandinavia, Estonia, and Lithuania, the southern sections of western Germany, Switzerland, Austria, the Czech Republic,

Slovakia, Hungary, and Poland. In eastern Germany it occurs in Thüringen and the Harz and Erz Mountains.

Biotope: In central Europe it is mostly found in foothills or at medium altitudes in mountains, open shrub and forest landscapes, as well as in towns. In mountain ranges it occurs to altitudes above 2290 meters (Alps), in winter at 2200 meters in caves. The highest known maternity roosts are at 1660 meters (Alps).

Summer roosts, as a rule, are located in cracks. Maternity roosts are frequently on or in houses roofed with slate or sheet metal, behind fireplace facades, shutters, and in cracks in the attic. Individual animals are also found in tree holes and fallen trees.

Winter roosts are located in caves, tunnels, and cellars. The temperature range is 1 to 5.5°C (7.5°C), and it can briefly (one to two days) tolerate temperatures as low as -5.5°C. It is found both free hanging on the wall or ceiling and wedged into cracks. Apparently it does not form clusters.

It hibernates from October to March or April, but this varies regionally.

Migrations: No detailed information is available, although probably it is more of a permanent resident. The longest migration was 115 kilometers.

Reproduction: Only sketchy information is available. The maternity roosts are occupied by approximately 20 to 60 females

starting in late April. Roosts are changed frequently.

Females give birth starting in the second half of June (Czech Republic and Slovakia). In northern regions and Russia there are usually one or two youngsters, in the Czech Republic and Slovakia primarily one. The young become capable of flight from mid to late July.

Maximum age: 14 years, 6 months.

Hunting and diet: This bat emerges from the roost in the early evening, sometimes not until after nightfall. Its flight is fast and agile with quick turns.

It hunts more in open country with free air space, over water, but also at treetop height. It hangs from branches during pauses in hunting. It also hunts in light rain and over street lamps.

It preys on flying insects.

Calls: Echolocation calls (searching flight) are FM calls from 38 to 28 kHz, merging into CF calls at 30 kHz. Their duration is 10 to 12 ms (see sonogram). The maximum impulse intensity is at 30 kHz. Calls are repeated at 200 ms intervals (= 5 calls per second). their range is up to 50 meters. There are similarities with the calls of the Parti-colored Bat and the Noctule (when hunting at low altitude).

Social calls are long bursts of 58 to 10 kHz that merge into CF components (up to 10 ms).

Protection: The Northern Bat is severely endangered in western Germany; rare in eastern

Germany; endangered in Austria.

Systematic protection of known maternity roosts, as with the Serotine, is necessary.

GENUS: *VESPERTILIO* LINNAEUS, 1758

The genus *Vespertilio* includes three species, one of which occurs in Europe.

$$\text{tooth formula} \quad \frac{2113}{3123} = 32$$

Vespertilio murinus Linnaeus, 1758
[= *Vespertilio discolor* (Natterer, 1818)]
Eng.: Parti-colored Bat
Ger.: *Zweifarbfledermaus*
Fr.: *Serotine Bicolor*
Head-body: 48-64 mm
Tail: 37-44.5 mm
Forearm: 40-47 (48.2) mm
Ear: 12-16.5 (18.8) mm
Wingspan: 270-310 mm
Condylobasal length: 13.9-15.7 (16.2) mm
Weight: 12-20.5 g
Distinguishing characters: Medium-sized. The ears are short, broad, and oval. The hind margin of the ear has four horizontal creases, runs with a broad fold to below the line of the corner of the mouth, and then rises toward it. The tragus is short, widening above, attaining its greatest width at about the second third of its length, and rounded at the tip.

The fur is long and dense. The roots of the hairs are blackish brown; on the back they have silvery white tips, giving them a frosted or "moldy" appearance. The underside is whitish gray and the throat is almost pure white, being sharply set off against the back. The ears, wing membranes, and muzzle are blackish brown. Juveniles are darker, more gray-black, the tips of the hairs a dirty gray-white; the belly is yellowish white. The wings are narrow and the arm membrane attaches at the base of the toe. The spur is longer than half the length of the tail membrane. A distinct epiblema with a visible cartilaginous steg is present. The last two tail vertebrae are free (3.5 to 5 millimeters). This is the only European bat with two pairs of milk teats, located 4 to 5 millimeters apart.

Anomalies in coloration: Unknown.

Similar species: The Northern Bat and Serotine are similar.

In the Barbastelle the coloration of the back is about the same, but the ear is completely differently shaped and the underside is dark.

Range: The parti-colored Bat is found in central and eastern Europe. In the north it occurs to about 60° latitude, often only through single reports. It is found in southern Sweden, southern Norway, is more abundant in Denmark, Estonia, western Germany (westernmost maternity roost in 1949 in Bavaria), single reports in eastern Germany, Switzerland, eastern France, and northern Italy. One maternity roost is known in the Czech Republic and Slovakia. It also occurs in the Balkan region.

In July of 1987, A. Hinkel and H. Zöllick found a large maternity roost on the Baltic Sea coast northwest of Rostock (eastern Germany).

Apparently this species breeds more in northern and eastern Europe and overwinters more to the west.

Biotope: Originally this was probably a cliff bat. It occurs in forested uplands and in plains, as well as on tall buildings in cities (substitute for cliffs?). In mountains it is found up to an altitude of 1923 meters (Col de Bretolet, Alps). Summer roosts are located primarily in cracks, such as behind shutters, in cracks in walls, in log cabins, and in the joists of attics. Winter roosts are located in caves, cellars, usually hidden in cracks, possibly also in tree holes. It hibernates from October to March.

Migrations: This is a migratory species. It migrates from the north and northeast to the west and southwest (winter roosts). In Belarus the bats migrate beginning in August over distances of up to 900 kilometers.

Reproduction: Only sketchy information is available.

The breeding season begins in August, at which time the testes of the males are greatly enlarged. Maternity roosts contain 30 to 50 females.

Females give birth in late June and early July to two, rarely three, young. Males form large colonies (up to more than 250 individuals) in the summer.

Maximum age: 5 years.

Hunting and diet: The Parti-colored Bat emerges from the

Parti-colored Bat.

Parti-colored Bat.

roost in late evening. Its flight is high (10 to 20 meters), fast, and straight. It hunts throughout the night. It preys on beetles and moths.

Calls: A variety of echolocation (searching) calls (see sonogram) are known.

I. An FM call of 50 to 20 kHz. The duration is 5 to 8 ms. the maximum impulse intensity is at 25 kHz. The calls are repeated at 180 to 200 ms intervals (= 5 to 6 calls per second).

II. A flat FM burst that falls off by only 2 kHz.

III. An undulating call lasting 40 ms.

Social calls include loud, shrill courtship calls in the fall. They last about 150 ms, beginning with 12 rapidly repeated FM calls (30 to 16 kHz). There then follows a burst (50 to 14 kHz) that turns into a CF call (approximately 10 ms) at 14 kHz. After that the frequency drops down to 10 kHz. Calls are repeated at 235 ms intervals (= 4.3 courtship calls per second).

Protection: This species is severely threatened in western Germany; while in eastern Germany it is a rare, endangered migrant; it is endangered in Austria. Because of a lack of information, protective measures are problematic. Protect known roosts.

PIPISTRELLUS KAUP, 1829

The genus *Pipistrellus* contains about 48 species, four of which occur in Europe. They have small, short ears, the tragus is rounded, the spur has an epiblema, and the muzzle, ears, and wing membranes are dark. The wings are relatively narrow. Two teats are present.

$$\text{tooth formula } \frac{2123}{3123} = 34$$

Pipistrellus pipistrellus (Schreber, 1774)
Eng.: Common Pipistrelle
Ger.: *Zwergfledermaus*
Fr.: *Pipistrelle Commune*
Head-body: (32) 36-51 mm
Tail: (20) 23-36 mm
Forearm: 28-34.6 mm
Ear: 9-13.5 mm
Wingspan: 180-240 mm
Condylobasal length: 11-11.8 mm
Fifth finger: 36-41 mm (males), 42 mm (females)
Weight: 3.5-8 g
Distinguishing characters: This is the smallest European bat. Its ears are short, triangular, and rounded at the tip. The outermost ear margin has four or five horizontal creases. The tragus is longer than wide, curves slightly inward, and is rounded on top.

The hair is dark brown to blackish brown at the base. The back is rufous, chestnut, or dark brown. The underside is yellow-brown to gray-brown. The muzzle, ears, and wing membranes are blackish brown. Juveniles are darker, more dark brown to blackish brown. The free margin of the wing membrane between the fifth finger and foot rarely have a narrow, indistinct, light edging. The wings are narrow, the arm membrane attaching at the base of the toe. The spur reaches about one-third of the length of the tail membrane. The epiblema is clearly developed and with a visible steg. The lower leg and the tail membrane are bare. The first premolar is small and displaced to the inside of the row of teeth. The second incisor is usually shorter than the small point of the first incisor. A glandular bump is located on the inside of the corner of the mouth. Animals disturbed in hibernation or daytime lethargy fall into fright paralysis (akinesis). Common Pipistrelles frequently stick feces to walls, window panes, and so forth.

Anomalies in coloration: Partial albinism (white skin and wing membranes, fur a lighter yellow-brown, eyes dark).

Similar species: Nathusius's Pipistrelle has the fifth finger longer, and the tail membrane is hairy toward the middle of the upper side, and on the underside along the lower leg.

Kuhl's and Savi's Pipistrelles are usually larger. Pay attention to the differences in coloration

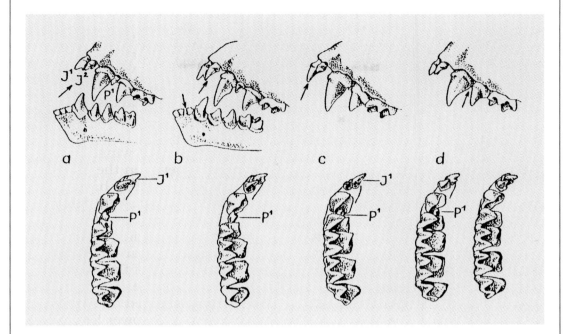

Dentition of the *Pipistrellus* species (viewed from the side and above).
a: Common Pipistrelle, b: Nathusius's Pipistrelle, c: Kuhl's Pipistrelle, d: Savi's Bat.
J: incisors, P1: first premolar in the upper mandible (for further details compare the species descriptions and the diagnostic key).

and tooth characters. In Savi's Pipistrelle the tragus is also differently shaped.

In the Whiskered Bat the ear and tragus are completely different.

Range: This species is distributed throughout almost all of Europe. In the north it is found to about 61° latitude, occurring in Ireland, England, and southern Scandinavia; in the south it extends to the Mediterranean and Balkan regions; in the east it reaches to the Caucasus.

Biotope: This is primarily a house bat. It occurs both in towns and in cities, as well as in parks and forests. Maternity roosts are usually below 600 meters. Individuals have been observed to a maximum altitude of 2000 meters.

Summer roosts (maternity roosts) are located in cracks accessible from the outside, in hollow spaces behind paneling, behind shutters, in half-timbered houses, as well as in shallow bat boxes. Individuals roost in the smallest cracks in walls and behind signs. It also inhabits suitable cracks in new buildings.

Winter roosts in northern and central Europe are found in large churches (up to 2000 individuals), in old limestone mines, in deep crevices in cliffs, in cracks in walls, and in cellars. In a cave in Romania, approximately 100,000 individuals were found in large clusters free-hanging from the ceiling. It is relatively insensitive to cold, living at temperatures of 2 to 6°C. It occasionally changes roosts even in winter. The hibernation phase lasts 1 to 4 weeks. It hibernates from mid November to early March or April.

Migrations: Most populations in central Europe are sedentary. The distances between the summer and winter roosts barely exceed 10 to 20 (50) kilometers. The longest migration of two females banded in the northern part of eastern Germany was 770 kilometers to the southwest and 540 kilometers to the southeast, respectively (migrations of more than 1160 kilometers from the Ukraine to southern Bulgaria cited frequently in the literature actually refer to Nathusius's Pipistrelle). Why individual animals undertake long migrations is not known.

Reproduction: Females and some males become sexually mature in the first year; though most males reach sexual maturity in the second year. Males occupy individual, stable territories inside the maternity roosts and defend them during the mating season (late August to late September) against other males. During the courtship flights they emit special social calls and give off a strong musky

Common Pipistrelle. The comparison with the human thumb makes it clear that this is one of the smallest European mammals.

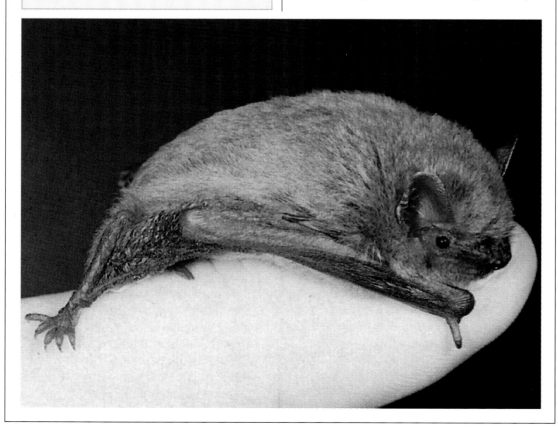

odor. The females occupy temporary mating roosts. A male has a harem of up to 10 females. Maternity roosts are occupied in April and May. Mixed maternity roosts with Nathusius's Pipistrelle are possible. The maternity roosts contain 20 to 250 (500) females.

Females give birth in June and early July. In central Europe the litter usually consists of two youngsters. The frequency of twin births increases from the west and south to the east and north. In England, as a rule there is only one youngster per litter. *Newborns*: Back pinkish, weight 1 to 1.4 grams, forearm length 11 to 12 millimeters. *Tenth day:* Forearm length 16 to 17 millimeters. *Twentieth day:* Forearm length 23-25.5 millimeters. *Thirtieth day:* Weight 4 to 4.3 grams, forearm length 27 to 28.6 millimeters, wingspan 170 to 180 millimeters. The eyes open at three to five days, and the permanent teeth are complete at four weeks. Juveniles are capable of flight at three to four weeks. Maternity roosts are abandoned by adult females in early August and by the juveniles in mid to late August. At this time so-called invasions occur in high, large spaces, also inside homes. Sometimes as many as 100 predominantly juvenile Common Pipistrelles (offspring from a maternity roost) take part (in search of a roost?).

Maximum age: 16 years, 7 months; average age 2 to 3 years.

Hunting and diet: This bat emerges from the roost early, sometimes before sunset, and in late fall even during the day. The flight is fast, agile, and 5 to 10 meters above the ground.

It hunts one to two kilometers from the roost over ponds, along forest edges, in gardens, and under street lamps, maintaining set flight paths. It preys on small moths, mosquitoes, and similar insects. It often flies back to the roost after only one to two hours.

Calls: A shrill scolding is emitted when it is disturbed.

Echolocation calls (searching calls) are of two types of calls (see sonogram).

I. The most common type is a short FM call of 80 to 58 kHz, 4 to 6 ms, that can end with a pure CF component. The maximum impulse intensity is at 58 kHz. Calls are repeated at 85 ms intervals (= 12 calls per second). In group flights the CF component can differ in frequency by about 14 kHz between individuals.

II. The second call is a pure CF call of 51 kHz of 10 ms duration. The calls are repeated at 95 ms intervals (= 10.5 calls per second).

The two types of calls are not used alternately as with the Noctule. The range is 20 to 50 ms.

A social call is an undulating, loud burst between 35 and 18 kHz, which rises and falls four times in 35 ms.

Protection: This species is endangered in western Germany and Austria. It is also endangered in eastern Germany, though there are still stable populations in many areas. In northern and central Europe this is still one of the most abundant species of bat.

Protection of the maternity roosts is necessary. Catch and release invading animals, and seal the entrance opening to an invaded roost. Inspect lamp shades, vases, pipes, and defective double-hung windows that are open on top ("bat traps"). Cats can catch bats at the roost. Put out bat boards or bat boxes and leave shutters on the windows.

PIPISTRELLUS NATHUSII (KEYSERLING ET BLASIUS, 1839)

Eng.: Nathusius's Pipistrelle
Ger.: *Rauhhautfledermaus*
Fr.: *Pipistrelle de Nathusius*
Head-body: 46-55 (58) mm
Tail: (30) 32.3-40 (44) mm
Forearm: 32-37 mm
Ear: 10-14 mm
Wingspan: 230-250 mm
Condylobasal length: (12.1) 12.3-13.2 mm
Fifth finger: (42) 43-48
Weight: 6-15.5 g

Distinguishing characters: Small species. The ears are short, triangular, and rounded at the tip. The outside margin of the ear has four or five horizontal creases. The tragus is short, curving slightly inward, and rounded at the tip.

The hair is dark brown at the base. In summer the back is rufous to chestnut brown, more dark brown after the molt (July through August), and often with a distinct gray suffusion. The underside is light brown to yellow

Common Pipistrelle. Hidden under the bulging wing membrane is a one-day-old youngster, of which only the foot and forearm are visible. Note that the foot of the youngster is nearly as large as that of the mother.

brown. The tail, ears, and wing membranes are blackish brown. Juveniles are dark brown without gray tones. The wings are long. The free margin of the wing membrane between the fifth finger and the foot often has a narrow, diffuse, yellow margin. The arm membrane attaches at the base of the toe. The tail membrane has hair on about half of the dorsal side, and only along the lower leg on the underside. The spur reaches about a third of the length of the tail membrane. The epiblema has a visible cartilaginous steg. The first premolar is well developed, visible from outside the mouth. The second incisor is longer than

the short point of the first incisor. A glandular bump is present on the inside corner of the mouth. As in the Common Pipistrelle, this species is also prone to fright paralysis (akinesis).

Anomalies in coloration: One individual with ashy gray fur has been documented.

Similar species: As in the Common Pipistrelle, please refer to the discussion of that species.

Range: this species occurs in eastern central and southern Europe. There are only two documented reports from England. In the north it is found in Denmark, southern Sweden, and along the Baltic Sea coast to the St. Petersburg region. In the south it occurs in the Mediterranean and Balkan regions. It also is found on Corsica and in the Caucasus.

Biotope: This is a forest bat. It occurs both in moist deciduous forests and in dry pine forests, as well as in parks. More rarely it is found in human settlements. It prefers lowlands. During migrations it is found at altitudes of up to 1923 meters (Col de Bretolet, Alps).

Summer roosts (maternity roosts) are located in tree holes, shallow bat boxes, cracks in tree trunks, cracks in hunting stands, and more rarely on buildings. It generally prefers cracks. Sometimes it shares maternity roosts with the Common Pipistrelle or Brandt's Bat.

Winter roosts are located in fissures in rocks, cracks in walls, caves, as well as in tree holes.

Migrations: This is a migratory species. Starting in mid August to September the entire population resident in the northern part of eastern Germany migrates to the southwest (France, Switzerland, western Germany). They appear again in April and May in the summer roosts. Migrations of more than 1000 kilometers (maximum of 1600 kilometers) have been documented repeatedly. The females migrate earlier in the fall than the males. So far no winter roosts have been found in Poland, eastern Germany, the Czech Republic, and Slovakia, but they have been reported in western Germany, Austria, and Switzerland.

Reproduction: Females become sexually mature in the first year, males not until the second year.

The mating season starts in the second half of July and runs to early September. The testes and epididymis are visibly enlarged. The sides of the muzzle are distended by glandular bumps. The males form breeding territories and harems of 3 to 10 females. The behavior in the breeding territory is similar to that of the Common Pipistrelle. The maternity roosts are occupied in April and May by 50 to 200 females. Young females often seek out the maternity roost in which they were born in the following year. The females are extremely site faithful, though they may change the maternity roost several times in the summer.

Females give birth in the second half of June, rarely earlier, to two young. *Newborns:* pink in color, weight 1.6 to 1.8 grams, forearm length 12 to 13.5 millimeters. *Tenth day:* weight 3.2 to 4.5 grams, forearm length 17.5 to 20 millimeters. *Twentieth day:* weight 5.2 grams, forearm length 28 millimeters. The eyes open on the third day. The young are capable of flight at about four weeks of age. Mother bats leave the maternity roost beginning in mid July and seek out the breeding roosts located up to 15 kilometers away.

Maximum age: 7 years.

Hunting and diet: This species emerges from the roost in early evening. Its flight is fast, in a straight-line flight in part with deep wing beats. It is not as agile in confined spaces as the Common Pipistrelle.

It hunts at altitudes of 4 to 15 meters in firebreaks, along paths and forest edges, as well as over water.

It preys on small to medium-sized flying insets.

Calls: When disturbed this bat emits a high-pitched, shrill scolding, like the Common Pipistrelle. A quiet whispering is often audible outside the occupied bat boxes.

The echolocation call (searching flight) is an FM call running into a CF component, 70 to 38 kHz, with a duration of 5 ms (see sonogram). Calls are

Nathusius's Pipistrelle in the rufous summer coat.

repeated at 100 to 125 ms intervals (approximately 8 to 10 calls per second).

Protection: This species is severely threatened in western Germany, endangered in Austria, and endangered in eastern Germany, though populations in the northern part of eastern Germany that have been studied for many years are stable. It is certainly more abundant in many areas than had previously been assumed, because occurrences can be documented only by working systematically with forest bats (bat boxes).

The installation of shallow boxes or bat boards on hunting stands has proven to be very successful.

> After the molt into the winter coat, Nathusius's Pipistrelle is more dark brown in color, in part with a slight gray tinge.

PIPISTRELLUS KUHLI (KUHL, 1819)

Eng.: Kuhl's Pipistrelle
Ger.: *Weissrandfledermaus*
Fr.: *Pipistrelle de Kuhl*
Head-body: 40-47 (48) mm
Tail: 30-34 mm
Forearm: 31-36 (37) mm
Ear: 12-13 mm
Wingspan: 210-220 mm
Condylobasal length: 12-13.2 mm
Weight: 5-10 g

Distinguishing characters: Small species. The ears are short, vaguely triangular, and rounded on top. The outside margin of the ear has five horizontal creases. The tragus is rounded, curving slightly inward, and does not broaden above.

The fur coloration is highly variable and dark brown at the base. The back is medium brown to yellow-brown, also pale cinnamon. The underside is light

gray to gray-white. The ears, wing membranes, and muzzle are dark brown to blackish brown.

Two views of the Kuhl's Pipistrelle.

The wings are relatively narrow. The arm membrane attaches at the base of the toe. The free margin of the wing membrane, particularly between the fifth finger and the foot, has a 1 to 2 millimeter wide white border, which is rarely absent. The spur has an epiblema that is divided by a visible steg. The first incisor has one point, the second incisor is very small, and the first premolar is displaced to the inside and is not visible from the outside. Glandular bumps are present at the inside corners of the mouth. Always use dental characters for identification.

Anomalies in coloration: A rare color variety is dark gray-brown to blackish brown on the back. The underside is only slightly lighter, the ears and wing membranes are nearly black, and the wing margins lack the white border.

Similar species: Savi's Pipistrelle has the tip of the tail free by 3 to 5 millimeters, its tragus is shorter and broadens above. The first incisor has two points.

In Nathusius's Pipistrelle and the Common Pipistrelle the first incisor has two points (for additional characters see those descriptions).

Range: This species occurs mainly in southern Europe: Portugal, Spain, southern France, Switzerland (breeding has been documented), Austria, Italy, Bulgaria, Croatia, Bosnia, Serbia, Montenegro, Greece, the Mediterranean islands, and to the Caucasus in the east.

Biotope: Kuhl's Pipistrelle occurs both in lowlands and at low altitudes in mountains. It is relatively closely tied to human settlements, but it also occurs in karst formations. It has been documented up to an altitude of 1923 meters (Col de Bretolet, Alps).

Summer roosts (maternity roosts) are located primarily in cracks on and in buildings (cracks in walls, openings under the roof). Individuals are also found in fissures in cliffs.

Winter roosts, as far as is known, are located in fissures in cliffs and cellars.

Migrations: Unknown, therefore this species apparently is a permanent resident.

Reproduction: Females become sexually mature in their first year. The small maternity roosts contain about 20 females. Females give birth to two young. Otherwise no detailed information is available.

It is similar to the Common Pipistrelle.

Maximum age: 8 years.

Hunting and diet: Kuhl's Pipistrelle emerges from the roost in late evening or in darkness.

It hunts at low or medium altitudes around street lamps, over water surfaces, and in gardens. Its flight is fast and agile.

It preys on small flying insects.

Protection: this species is endangered in Austria. It is hardly possible to estimate the degree of endangerment in the main area of occurrence. The systematic protection of known summer roosts is necessary.

PIPISTRELLUS SAVII (BONAPARTE, 1837)

[*Hypsugo savii* (Kolenati, 1856)]
Eng.: Savi's Pipistrelle
Ger.: *Alpenfledermaus*
Fr.: *Pipistrelle de Savi*
Head-body: 40-54 mm
Tail: 31-42.5 mm
Forearm: 30-36.5 (38) mm
Ear: (10) 12-15 mm

its morphological characters it is intermediate between the genera *Eptesicus* and *Pipistrellus*. This is a small species. The ears are broader and rounder than in the other European *Pipistrellus* species. The outside margin of the ear has four horizontal creases. The tragus is short, broadening slightly above, and

Savi's Pipistrelle.

Wingspan: 220-225 mm
Condylobasal length: 11.9-13.6 (14) mm
Weight: 5-10 g
Distinguishing characters:
Horácek and Hanák consider this species to belong in its own genus, *Hypsugo,* separate from the genus *Pipistrellus*. Based on

the length of its inside margin is nearly the same as its greatest width. At the base of the outside margin there are two overlapping "teeth." The rounded tip of the tragus curves inward.

The fur is relatively long and blackish brown at the base. The back varies from pale yellow-brown to dark brown with gleaming golden tips. The underside is light whitish yellow

Savi's Pipistrelle.

to grayish white, in clear contrast to the back. The ears and muzzle are blackish brown or black, in clear contrast to the remaining coloration. The wing membranes are dark brown. The arm membrane attaches at the base of the toe. The spur has a narrow epiblema, but no visible steg. The last one or two tail vertebrae are free (3 to 5 millimeters). The first incisor has two points. The first premolar is not visible from outside. It is displaced to the inside of the row of teeth and is completely absent in 10 to 40% of the individuals. This is a very lively species behaviorally.

Anomalies in coloration: Unknown.

Similar species: Savi's Pipistrelle is similar to the Common, Kuhl's, and Nathusius's Pipistrelles. Please refer to these species for comparisons.

The Northern Bat is similar in coloration, but larger.

Range: This species is found primarily in southern Europe, making the German name (Alpine Bat) misleading. It occurs in Spain, southern France, Bulgaria, Croatia, Bosnia, Montenegro, and Greece. there have been no reports in recent decades from Germany and Austria. There is a documented maternity roost in Switzerland.

Biotope: Savi's Pipistrelle inhabits mountain valleys, alpine pastures, and karst formations, as well as in human settlements. On Mediterranean islands it is found directly on the coast. It has been documented in mountains to an altitude of 2600 meters.

Summer roosts (maternity roosts) are often located in cracks in and on buildings (rafters, cracks in masonry, hollow spaces between bricks, holes in walls) and crevices in rock faces.

Winter roosts are located in lower lying valleys in caves and crevices in cliffs and probably also in tree holes.

Migrations: Because of the lack of field studies, its migratory status cannot be determined for certain. Apparently it is a partial migrant. The longest migration is over 250 kilometers.

Reproduction: Information on its reproductive biology is very sketchy.

The mating season begins approximately in late August and September. Maternity roosts contain 20 to 70 females.

Females give birth to two young from mid June to early July.

Maximum age: Unknown.

Hunting and diet: This bat emerges from the roost shortly after sundown. Its flight is straight, steady, and not very fast; in part it flies above houses and the treetops.

It hunts almost all night. On islands in the Adriatic, animals were also observed flying just above the ocean's surface during the day from seaside cliffs.

It preys on small flying insects.

Protection: This species is classified as severely endangered in western Germany, and in Austria as extirpated or absent. Precise estimates are impossible because of sketchy information.

LECOTUS GEOFFROY, 1818

The genus *Plecotus* contains five species, two of which occur in Europe. The ears are over 30 millimeters long and are connected in front at the base by a fold of skin. The nostrils open above. Echolocation calls can be sent out with the mouth closed and also through the nose. The spur lacks an epiblema.

$$\text{tooth formula } \frac{2123}{3133} = 36$$

Plecotus auritus (Linnaeus, 1758)
Eng.: Common Long-eared Bat
Ger.: *Braunes Langohr*
Fr.: *Oreillard Septentrional*
Head-body: 42-53 (55.5) mm
Tail: (32.5) 37-55 mm
Forearm: (35) 37-42 mm
Ear: 31-41 (43) mm
Wingspan: 240-285 mm

Condylobasal length: (13.2) 14-15.6 mm

Weight: 4.6-11.3 g

Distinguishing characters: Medium-sized species. The ears are conspicuously long. The external ear is thin and has 22 to 24 horizontal creases. The forward margin of the ear is broadened and has ciliate hairs. A button-like process is located near the base. The ears are erected only just before taking off and in flight. At other times they are folded and tilted backward (calling to mind ram's horns). In daytime lethargy and in hibernation the ears are folded back and tucked under the wings. The long, lanceolate tragus curves forward even when the ear is folded. The eyes are relatively large. The muzzle is swollen on the sides.

The fur is long and sparse and dark gray-brown at the base. The back is light gray-brown. At the boundary with the underside there is a generally lighter, yellowish-brown patch on the side of the neck. The underside is light gray, sometimes with a yellowish tinge. The lips are light flesh-colored. The nose and eye regions are light brown; the ears and wing membranes are light gray-brown. The tragus is yellowish white, with a slight light gray pigmentation toward

Common Long-eared Bat (left) and gray Long-eared Bat (right). Clearly visible are the differences in the fur and facial coloration as well as the length of the thumb and claw. Both animals are in fright posture, in which they cower and lay their ears flat.

the tip. Juveniles are pale gray without brown tones and have a dark face. The wings are broad, the arm membrane attaches at the base of the toe, and the spur attains about half the length of the tail membrane. The penis is thin and tapers at the end. The feet are large and the thumbs and thumb claws are long.

Thumbs: >6 (6.5-8.4) millimeters

Thumb claws: (1.5) 2.5-3 millimeters, note the amount of wear!

Tragus width: < 5.5 (4.5-5.2) millimeters

Foot: 6.5-9.2 (11) millimeters

Anomalies in coloration: Partial albinism (wing membranes and parts of the ears).

Similar species: The Gray Long-eared Bat has the fur dark at the base, slate gray, while the back appears altogether gray. The muzzle, upper lip, and tragus are gray. Note the measurements of the thumbs, thumb claws, tragus, and feet.

In Bechstein's Bat the ears are shorter and clearly separated from each other at the base.

Range: This species occurs throughout almost all of Europe, including northern Ireland and England. In Scandinavia it occurs to about 64° latitude. It has not been documented in southern Spain, southern Italy, or Greece. It is also present in the Caucasus.

Biotope: The Common Long-eared Bat prefers open deciduous and coniferous forests in lowlands and mountains at medium altitude. It also occurs in parks and gardens in towns and cities, but is not bound to settled regions. The highest maternity roost was found at 1660 meters (Switzerland), otherwise it ranges to an altitude of 2000 meters, though it usually stays at lower elevations.

Summer roosts (maternity roosts) are located in tree holes, bat boxes and bird houses, in attics, and sporadically in holes in cliffs, behind shutters, and in cracks in buildings.

Winter roosts are found in cellars, tunnels, and caves, rarely in thick-walled tree holes. Temperatures range from 2 to 5°C, but it can also tolerate temperature drops down to -3.5°C for one to two days. As a relatively hardy species, it is usually found closer to the cave entrance than *Myotis* species. In the roost it wedges into cracks, as well as deep in narrow pipes, and sometimes also free hanging on the wall. The wing membranes partially cloak the belly and chest. Usually it roosts solitarily, rarely in small clusters (two or three individuals). It also forms mixed roosts with other species.

It hibernates from October and November to late March and early April.

Migrations: This species is a permanent resident, usually traveling only a few kilometers between the summer and winter roosts. The longest migration is 42 kilometers.

Reproduction: Females become sexually mature in their second year.

Common Long-eared Bat.

The mating season is in the fall (until spring?). Maternity roosts are occupied starting in April and May by 10 to 50 (100) females. Males are solitary in the summer and only rarely turn up in the maternity roosts. Maternity roosts are located in attics, seldom in one spot. The females usually hang scattered on the ridge or occupy cracks.

Females give birth to one, rarely two, young starting in mid June. The eyes open on the sixth day. The ears become erect on about the eleventh day. Juveniles are capable of flight in mid to late July. In the fall, invasions of up to 10 individuals in homes are possible (juveniles searching for roosts).

Maximum age: 22 years. The average age is 4 years.

Hunting and diet: This bat emerges from the roost in late evening, though usually not until after dark. Its flight is slow and fluttering, but it can hover and is very agile in the most confined spaces.

It preys on moths (primarily noctuids). Harvests caterpillars, spiders, and other prey are taken from twigs and other substrates. In this way it also preys on butterflies (Peacock Butterfly and

Tortoiseshell Butterfly among others). The prey is consumed in part at feeding roosts, where wing remains accumulate. Observations in captivity indicate the supplemental use of vision to detect prey animals.

Calls: For defense it produces a relatively low-pitched chirping or humming sound. In spring and fall it emits a "tzick-tzick" call in flight.

keeping with their hunting behavior (hover by or circle around leaves and twigs). Occasionally they also produce a loud burst from 42 to 12 kHz, with a duration of 7 ms. Calls are repeated at 180 to 200 ms intervals (= 5.4 calls per second).

Protection: This species is severely endangered in western Germany; endangered in Austria; in eastern Germany, however, it

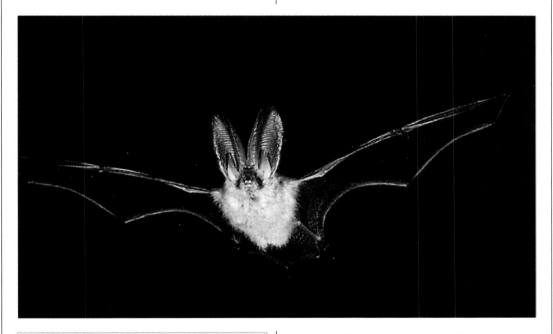

Gray Long-eared Bat in flight. Because Long-eared Bats can also send echolocation calls through the nostrils, the mouth is closed.

Echolocation calls (searching flight) are soft, short FM calls of 83 to 26 kHz with a duration of 2 ms (see sonogram). The maximum impulse intensities are at 26, 42, and 59 kHz. Calls are repeated at 50 ms intervals (= 20 calls per second). They have a short range of up to 5 meters. Long-eared Bats whisper, in

is only threatened locally.

The systematic protection of roosts and the biotope and putting out bat boxes are necessary.

***PLECOTUS AUSTRIACUS* (FISCHER, 1829)**
 Eng.: Gray Long-eared Bat
 Ger.: Graues Langohr
 Fr.: Oreillard Méridional
 Head-body: 41-58 (60) mm
 Tail: 37-55 (57) mm
 Forearm: (35) 37-44.5 mm

Gray Long-eared Bat. In comparison to most other European common bats, long-eared bats have relatively large eyes.

Ear 31-41 mm
Wingspan: 255-292 mm
Condylobasal length: 15-17 (17.2) mm
This species was not rediscovered again in central

muzzle is longer and more pointed and the eyes are relatively large.

The fur is long, dark slate gray at the base. The back appears predominantly gray, with at most a slight brownish tinge. The underside is light gray. The muzzle and upper lip are dark gray. A gray mask is present, particularly around the eyes. The wing membranes and ears are gray. The tragus has a gray pigmentation almost from the base on. The wings are broad. The arm membrane attaches at the base of the toe. The spur is nearly half the length of the tail membrane. The tip of the penis is thickened. The thumbs, thumb claws, and feet are small.

Thumb: <6 millimeters
Thumb claw: <2 millimeters
Foot: 6-8 millimeters
Tragus width: >5.5 millimeters
Weight: 5-13 grams
figure, page 171

Anomalies in coloration: Unknown.

Similar species: Common Long-eared Bat is the most similar species. It also resembles slightly Bechstein's Bat (see that species for characters).

Range: The Gray Long-eared Bat occurs in central and southern Europe. In the north it is found in southern England, France, and locally in Belgium and southern Holland. In Germany and Poland it occurs to about 53° latitude and does not reach the coast of the Baltic Sea. It is abundant in the Mediterranean and Balkan regions and in the Caucasus.

Europe until 1960 by Bauer.

Distinguishing characters: Medium-sized species. The ears are as in the Common Long-eared Bat, with approximately 22 to 24 horizontal creases, the

Biotope: This is a warmth loving bat. It prefers cultivated landscapes, in low mountains in warm valleys, usually below 400 meters altitude. In the north it is largely tied to human settlements (a house bat) and avoids extensive forests. In the summer it has been documented up to an altitude of 1380 meters (Switzerland), in winter up to 1100 meters. Summer roosts (maternity roosts) are located in buildings, in part free hanging on the ridge, in part hidden in cracks and mortises in beams. Sometimes it shares roosts with the Greater Mouse-eared Bat and Lesser Horseshoe Bat. Individual animals are also found in caves.

Winter roosts are located in caves, cellars, tunnels, in part together with the Common Long-eared Bat. Temperatures range from 2 to 9°C (12°C). It roosts free hanging on the wall more frequently than the Common Long-eared Bat, but it is also found in cracks; usually alone, rarely two or three animals together.

It hibernates from September or October to March or April.

Migrations: This species is a permanent resident. The distance between the summer and winter roosts is less than 20 kilometers. The longest migration is 62 kilometers.

Reproduction: Information is sketchy. It forms small maternity roosts, usually with only 10 to 30 females. It roosts free-hanging, sometimes in small groups, more or less hidden in attics. No maternity roosts have been found so far in tree holes or bat boxes.

Females give birth to one youngster in mid to late June.

Maximum age: 14 years, 6 months.

Hunting and diet: It emerges from the roost after dark. Its flight is similar to that of the Common Long-eared Bat. It is very agile.

It often hunts in the open air, as well as around street lights.

It primarily preys on moths, dipterans, and small beetles. The harvesting of food animals from branches has yet to be confirmed. It uses feeding roosts.

Calls: It makes chirping or humming sounds when disturbed.

Protection: This species is severely endangered in western Germany, and endangered in Austria and eastern Germany. It is rarer than the Common Long-eared Bat.

Protection of the maternity roosts (use of wood preservatives that are not toxic to mammals) and of the biotope are necessary.

BARBASTELLA GRAY, 1821

The genus *Barbastella* includes two species, one of which occurs in Europe.

$$\text{tooth formula} \quad \frac{2123}{3123} = 34$$

Barbastella barbastellus (Schreber, 1774)
Eng.: Barbastelle
Ger.: Mopsfledermaus

Fr.: Barbastelle
Head-body: 45-58 mm
Tail: (36) 38-52 mm

Barbastelle. With this species, small orange mites often attach themselves to the outside margin of the ear.

Forearm 36.5 - 43.5 mm
Ear: 12.1-18 mm
Wingspan: 262-292 mm
Condylobasal length: 12-14.7 mm
Weight: 6-13.5 g

Distinguishing characters:
Medium-sized species. The muzzle has a compressed, pug-nosed appearance; the nostrils open on top. The external ear is broad and opens to the front; the inside margins of the ears are fused in the middle at the base. The outside ear margin has 5 or 6 horizontal creases. Approximately in the middle there is a conspicuous, button-like lobe of skin that appears as if stuck on, but which also can be completely absent (for example, it is fully developed in only about a third of the animals in the Czech Republic and Slovakia; in Poland, Lithuania, and eastern Germany it is present in more than 90% of all individuals). The tragus is triangular, with a long, tapering, rounded tip. The eyes are small, the mouth opening very narrow, and the teeth are small.

The fur is long, silky, and black at the base. The back appears blackish brown with whitish or yellowish white tips, as if frosted. The underside is dark gray, with the bare parts of the face and the ears black, and the wing membranes are gray-brown to blackish brown. The wings are narrow and long. The arm membrane attaches at the base of the toe. The spur reaches about half the length of the tail membrane. A narrow epiblema with a cartilaginous steg is present. One pair of teats is present. Juveniles are somewhat darker when they are capable of flight, but the hairs on the back already have whitish tips.

Anomalies in coloration:
Partial albinism is common (Czech Republic and Slovakia).

Similar species: This species is unmistakable in central Europe. In the Caucasus, the very similar, somewhat larger (forearm length 41.2 to 45 millimeters, condylobasal length 14.2 to 14.9 millimeters) Asiatic species *Barbastella leucomelas* is supposed to occur sympatrically with *Barbastella barbastellus*. Contrary to the opinion of Kusjakin and other authors, the absence of the fold of skin on the outside margin of the ear is not useful for distinguishing the two species, because it also occurs in *B. barbastellus*. The status of *B. leucomelas* as an independent species is questioned by many authors. They consider *B. leucomelas* merely to be a subspecies of *B. barbastellus*.

Range: This species occurs in Europe from southern England to the Caucasus. In Norway and Sweden it is found to approximately 60° latitude. It is not abundant in most regions and has not been documented in parts of Spain, Italy, and the Balkans.

Biotope: It prefers forested foothills and mountains, but also occurs in towns. It has been documented in the summer at altitudes of up to 1923 meters (Col de Bretolet, Alps); the

highest maternity roost is at 1100 meters in the Czech Republic.

Summer roosts (maternity roosts) are located in cracks in buildings and frequently behind shutters. Individuals also roost in tree holes, nest boxes, or the entrances of caves.

Winter roosts are located in caves, tunnels, and cellars. A hardy species, it exists at temperatures between 2 to 5°C, and rarely to -3°C or lower. In the roost it is often found near the entrance, both in crevices and free hanging on the wall or

Barbastelle.

ceiling, occasionally in large clusters. Large winter roosts (western Poland) contain more than 1000 individuals. In many roosts there is a conspicuous predominance of males.

It hibernates from October and November to March and April.

Migrations: This is a partial migrant. Its longest migrations are up to 300 kilometers, but are usually shorter.

Reproduction: Females reach sexual maturity in their second year. Mating usually takes places in the fall, but also in the winter roosts. Maternity roosts often contain only 10 to 20 females, rarely up to 100. During this time males live in small groups outside the maternity roosts. This species is very sensitive to disturbance!

Females give birth starting in mid June, usually to two young.

Maximum age: 23 years.

Hunting and diet: This bat emerges from the roost in early evening. Its flight is fast and agile.

It hunts at treetop height along forest edges, in gardens, and along avenues.

It preys on small, delicate insects (moths, dipterans, small beetles). It cannot manage larger insects that have a hard exoskeleton because of its narrow mouth opening and weak jaws.

Calls: When disturbed, it produces a high-pitched chirping, as well as a muffled humming.

Two types of echolocation calls (searching flight) are produced (see sonogram):

I. A loud, short CF/FM call of 35 to 28 kHz, with a duration of 4 ms.

II. A soft, short CF/FM call of 43 to 33 kHz, with a duration of 5.2 ms. Both calls begin with a CF component of 1 to 1.5 ms duration. The maximum impulse intensity is at 35 to 30 and 43 kHz, respectively. Calls are repeated at 110 to 120 ms intervals (= 8 to 9 calls per second).

Protection: This species is threatened with extirpation in western Germany. It is endangered in Austria and severely endangered in eastern Germany. The causes for the in part extreme declines in population are not precisely known, but the drastic deterioration of the food supply through the use of insecticides is a possibility.

The systematic protection of the biotope, particularly of the known maternity and winter roosts, is necessary.

MINIOPTERUS BONAPARTE, 1837

The genus *Miniopterus* contains ten species, one of which occurs in Europe

$$\text{tooth formula} \quad \frac{212(3)}{3/3133} = 36\ (38)$$

Miniopterus schreibersi (Kuhl, 1819)

Eng.: Schreiber's Bat
Ger.: Langflügelfledermaus
Fr.: Minioptére
Head-body: (48) 50-62 mm
Tail: (47) 56-64 mm
Forearm: (44) 45.4-48 mm
Ear: 10-13.5 mm
Wingspan: 305-342 mm
Condylobasal length: 14.5-15.5 mm
Weight: 9-16 g

Distinguishing characters: Medium-sized. This species has a very short muzzle and a forehead hump. The ears are short, triangular, spaced far apart, and do not extend over the top of the head. Four or five horizontal creases are present. The tragus is short, curves inward, and is

Schreiber's Bat. Note the cinnamon brown chin patch and the lightened tragus in this adult animal (subspecies: *Miniopterus schreibersi inexpectatus*) from the Rhodope Mountains (Bulgaria).

rounded at the tip.

The fur on the head is short, dense, and erect. The back is gray-brown to ashy gray, in part with a pale mauve tinge. The underside is a somewhat lighter gray. Adult specimens in Romania and Bulgaria have a contrasting yellow to cinnamon-colored throat patch, as well as hairs on the forehead of a similar color. This coloration is absent in the totally gray (without shades of brown) juveniles. (Subspecies *Miniopterus schreibersi inexpectatus* Heinrich, 1936.) The muzzle, ears, and wing membranes are gray-brown, the tragus is yellowish white, partly with a light gray pigmentation (juveniles?). The wings are long and narrow. The second phalanx of the third finger is about three

Schreiber's Bat.

times as long as the first phalanx. At rest the third and fourth fingers are tucked at the phalanx to the inside between the first and second joints. This position is usually even maintained with passively spread wings (the other common bats bend all the phalanges of the third to fifth fingers at the basal joint with folded wings). The arm membrane attaches at the ankle. The feet are relatively long, as is the tail. The spur attains about one-third to one-half the length of the tail membrane. The epiblema is absent. This species has a very calm temperament.

Anomalies in coloration: Albinism.

Similar species: This species is unmistakable in Europe.

Range: It occurs in southern and eastern Europe. In the north it reaches Spain, France, Switzerland, Austria, Slovakia, and Hungary. It is widely distributed in the Mediterranean and Balkan regions, and is also found in the Caucasus.

Biotope: This is a cave bat. It occurs both in the lowlands and in the mountains (up to an altitude of 1000 meters). It is also found in karst formations.

Summer roosts (maternity roosts) are located in caves and occasionally in large attics in the northern part of the range.

Winter roosts are located in caves with a temperature of 7 to 12°C. The bats hang freely on the wall or ceiling, sometimes in clusters. Winter roosts can even be changed in the winter.

The hibernation period extends from October (?) to late March.

Migrations: This is a migratory species, at least in the north. Winter roosts are located up to 100 kilometers and more farther south than are the summer roosts. The longest migration is 350 kilometers.

Reproduction: Females attain sexual maturity in their second year. Mating takes place in the fall. In contrast to all other European bats, fertilization is immediate, but embryonic development is suspended during hibernation and does not start up again until the spring. The gestation period, therefore, is 8 to 9 months.

Maternity roosts often contain more than 1000 females; males are often present in the maternity roosts as well.

Females give birth in late June and early July to one (rarely two) youngster; the females are also said to nurse unrelated young.

Maximum age: 16 years.

Hunting and diet: This bat emerges from the roost shortly after sundown. Its flight is very fast (50 to 55 kilometers per hour) and resembling that of swallows or swifts.

It hunts at an altitude of 10 to

20 meters in open countryside, often far from the roost.

It preys on moths, mosquitoes, and beetles.

Calls: The defensive call is a shrill, short scolding. At rest in a group it makes a low-pitched, whispering call, resembling the contact calls of goslings or ducklings.

Protection: This species is classified as an endangered summer visitor. It is threatened with extirpation in Austria. Apparently it is a very sensitive species. Large colonies have died out in France and Switzerland.

The systematic protection of roosts is necessary.

Free-tailed Bats, Family Molossidae

TADARIDA RAFINESQUE, 1814

This species contains fifty-two species, one of which occurs in Europe.

$$\text{tooth formula} \quad \frac{1123}{3123} = 32$$

Tadaria teniotis (Rafinesque, 1814)

Eng.: European Free-tailed Bat

Ger.: Europäische Bulldoggfledermaus

Fr.: Molosse de Cestoni

Head-body: 81-92 mm

Tail: 44-57 mm

Forearm 57-64 mm

Ear: 27-31 mm

Wingspan approximately: 410 mm

Condylobasal length: 20.9-24 mm

Weight: 25-50 g

Distinguishing characters: Very large species. The ears are long, broad, and projecting forward over the eyes and face. The ears touch in front at the base; the hind margin of the ear is broadened at about eye height, where it is nearly rectangular. A conspicuous fold of skin (antitragus) is present. The muzzle is long, the upper lip has five creases, the nostrils open forward, and the eyes are large.

The fur is short, fine, and soft, almost like moleskin. The back is blackish gray to smoky gray, with a brownish tinge. The underside is a somewhat lighter gray. The ears, muzzle, and wing membranes are blackish gray. Juveniles are altogether more gray in color. The tail membrane is short. Up to one-third to one-half of the tail extends freely beyond the tail membrane. The spurs lack an epiblema. The wings are very long and narrow and the arm membrane attaches at the ankle. They can run very well and climb in crevices. The legs are short and robust. These bats smell intensely of a mixture of musk and lavender.

Anomalies in coloration: Unknown.

Similar species: This species is unmistakable in Europe.

Range: It occurs in southern Europe, in the Mediterranean region. It has been documented (often only through single records) in Portugal, Spain, southern France, Switzerland,

Italy, Bulgaria, Croatia, Bosnia, Serbia, Montenegro, Greece, and on islands in the Mediterranean.

Biotope: This is a cliff bat. It occurs both in mountains with steep cliff faces and gorges and in towns. It has been documented up to an altitude of 1923 meters (Col de Bretolet, Alps).

Summer roosts are located in crevices in cliffs, on Mediterranean islands also in crevices in riparian cliffs, on ceilings of caves, and in cracks in buildings.

Winter roosts are unknown. It is not clear whether a fairly long hibernation period is observed.

Migrations: Apparently this species is a partial migrant or a migrant.

Reproduction: Insufficient information is available on the reproductive biology of this species. Females apparently become sexually mature at one year of age.

Females give birth to one youngster, which becomes independent after 6 to 7 weeks.

Maximum age: The maximum age is not precisely known, but it is more than 10 years.

Hunting and diet: Sometimes it emerges from the roost in early evening, but usually not until later. Its flight is high, fast, and direct. It requires open air space. Sometimes it also hunts over bodies of water, where it flies in circles.

It preys on flying insects.

Calls: In flight it produces a loud, sharp "tzick." It also makes whistling sounds.

Echolocation calls (searching flight) are CF calls, largely of constant frequency, as a sustained tone of approximately 20 ms duration. There is a slight drop in frequency from 18 to 10 kHz (see sonogram). Calls apparently are repeated at variable intervals (1 to 4 calls per second).

Protection: Because there is only sketchy information on the habits, no specific advice is possible.

Mouse-eared Bat colony.

DIAGNOSTIC KEY TO THE EUROPEAN BATS

DIAGNOSTIC KEY TO THE EUROPEAN BATS

This key is based on the external characters of living, adult bats and makes it possible to identify all the European species, separated by family. Only a magnifying glass and a ruler (preferably calipers) are necessary. In any case, you can also refer to the figures in the text. The sequence of the characters given corresponds approximately to their reliability for identification.

Young bats (note the epiphyses, see figure) are usually darker and duller in coloration. When in doubt they should be identified by a specialist.

KEY TO THE FAMILIES

A. Nose with cutaneous processes (Horseshoe Bat, Saddle Bat, Lancet Bat, plate 1); ears lacking tragus
Horseshoe Bats (Family Rhinolophidae)

B. Nose lacking cutaneous process; ears with tragus (see plate 3); tail included completely or except for last two vertebrae (approximately 4 to 5 mm) in tail membrane (see plate 5)
Common Bats (Family Vespertilionidae)

C. Tail extending up to a third to a half beyond tail membrane; lower hind ear margin (antitragus) with pronounced lobe formations (plate 2i); no nasal processes

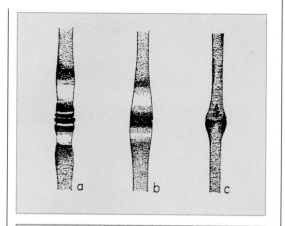

Degree of ossification of the epiphyses of the bones of the metacarpus and digits in (a) youngsters, (b) half-grown juveniles, and (c) adult bats (from Gromov, 1963).

Free-tailed Bats (Family Molossidae)

KEY A: HORSESHOE BATS

1a. Upper saddle process blunt in profile (plate 1b, 1d)

1b. Upper saddle in profile pointed and always longer than the lower saddle process (plate 1f, 1h, 1j)

2a. Forearm shorter than 50 mm, nasal process as in plate 1a, 1b Greater Horseshoe Bat (*Rhinolophus ferrumequinum*).

2b. Forearm shorter than 43 mm, nasal process as in plate 1c, 1d Lesser Horseshoe Bat (*Rhinolophus hipposideros*).

3a. Second phalanx of fourth finger more than twice as long as the first (plate 6k); lower saddle process broad when viewed from front, rounded; horizontal furrow under lancet without indentation in middle when viewed from the front (plate 1e, 1g)

3b. Second phalanx of fourth finger at most twice as long as the first; lower saddle process narrow when viewed from the front, rounded; horizontal furrow under lancet slightly indented in middle when viewed from the front (plate 1i); lancet narrows more or less uniformly above (plate 1I); saddle in profile as in plate 1j. Blasius's Horseshoe Bat (*Rhinolophus blasii*).

4a. Lancet narrows above more or less uniformly (plate 1e); upper saddle process pointed in profile, forward curving, clearly longer than lower process (plate 1f)... Mediterranean Horseshoe Bat (*Rhinolophus euryale*).

4b. Lancet suddenly narrows above starting in middle and tapers to sharp point (plate 1g); upper saddle process relatively blunt in profile and only slightly longer than lower process (plate 1h); a few conspicuously dark hairs often present around eyes ("spectacles") Mehely's Horseshoe Bat (*Rhinolophus mehelyi*).

KEY B: COMMON BATS

1a. Ears connected in front at base by fold of skin (plate 2g, 4i); nostrils open above (plate 2g, 2h; 4h, 4I)... Genera *Barbastella* and *Plecotus*

1b. Ears widely separated in front (plate 3); nostrils open to the front (plate 3)
     Genera *Myotis*, *Miniopterus*, *Eptesicus*, *Pipistrellus*, *Nyctalus*, and *Vespertilio*

2a. Ears over 30 mm long, numerous horizontal furrows, folded at rest

2b. Ears short and wide, a button-like process in middle of outside margin, which may also be absent (plate 4h, 4i); dorsal fur blackish brown with light tips, appearing as if frosted; spur with epiblema, divided by steg (plate 6a)...Barbastelle (*Barbastellus barbastellus*).

3a. Tragus with gray pigmentation nearly from base on, over 5.5 mm wide (plate 2h); thumb less than 6 mm long, thumb claw short, about 2 mm long (plate 7h); face with dark mask surrounding eyes; facial hair with gray pigmentation (plate 2h); dorsal fur predominantly gray, dorsal hairs dark slate gray at base; underside gray white, without brownish tinge; tip of penis thickened (plate 7d)...Gray Long-eared Bat (*Plecotus austriacus*).

3b. Tragus light, weakly pigmented toward tip, less than 5.5 mm wide (plate 2g); thumb over 6 mm long, claw long and pointed, 2.5 to 3 mm long (plate 7g); facial skin more brownish flesh-colored, only dark patch of pigment around eyes, no dark mask (plate 2g); dorsal fur more gray-brown, dorsal hairs dark brown at the base, underside gray white with brownish tinge; tip of penis not thickened (plate 7c)...

Common Long-eared Bat (*Plecotus auritus*).

4a. Spur with epiplema (plate 6c); tragus short, curved, rounded at tip (plate 4a) or with mushroom-shaped broadening (plate 2e)...Genera *Nyctalus, Pipistrellus, Eptesicus, Vespertilio*

4b. Spur without epiplema (plate 5e); when narrow edging of skin is present (plate 5c), then tragus tapering and lanceolate, reaching about half the ear length or projecting beyond it (for example, plate 3g) Genera *Myotis, Miniopterus*

5a. Last tail vertebra extending maximum of 1 mm beyond tail membrane (plate 5g); epiplema broad, with visible steg

5b. Last 1 or 2 tail vertebrae extending beyond tail membrane by 4 to 5 mm; epiplema narrow, usually without visible steg (plate 6b)

6a. Forearm longer than 47 mm; dorsal fur dark brown, in part shiny, but without golden sheen, underside yellowish brown, no sharp boundary between back and underside; epiplema narrow, without visible steg, tail membrane (plate 6b); ear with narrow hind margin, ending before corner of mouth (plate 4f)...Serotine (*Eptesicus serotinus*).

6b. Forearm at most 47 mm long; dorsal fur in adults with light-yellow or white tips.

7a. Hind ear margin with broad furrow extending down below line of corner of mouth and ending by it (plate 4g);

epiplema broad, with visible steg; dorsal hairs dark brown at base, appearing frosted because of white hair tips, underside whitish with sharp boundary toward dorsal side; females with 4 teats; forearm 40 to 47 mm

Parti-colored Bat (*Vespertilio murinus*).

7b. Hind ear margin with narrow furrow, extending in direction of corner of mouth and ending before it (plate 4d, 4e); epiplema without visible steg; dorsal hairs dark brown at base, tips yellowish or with golden sheen.

8a. Tragus clearly longer than wide, slightly curved (plate 4e); epiplema narrow; dorsal fur with golden sheen, underside yellowish brown, with sharper boundary to dorsal side only at throat; forearm 38 to 44 mm Northern Bat (*Eptesicus nilssoni*).

8b. Tragus short, broadening slightly above (plate 4d), length of its front margin corresponding nearly to its greatest width, two superimposed serrations at the base of the outside margin; ears and face black, contrasting strongly with light brownish yellow tips of dorsal fur and nearly white underside; forearm 30 to 38 mm...Savi's Bat (*Pipistrellus savii*).

9a. Tragus broadens above into mushroom shape (plate 2d, 2e, 2f); forearm longer than 38 mm.

9b. Tragus not broadened above into mushroom shape; forearm shorter than 38 mm

10a. Teeth: I1 as a rule with single point, I2 very small, P1 not visible from outside, displaced to the inside (see figure, page 149); sharply set off, approximately 2 mm wide, white stripe on margin of arm membrane between 5th finger and leg (caution: dark gray brown individuals without white wing-membrane margin also occur!); ears brown (plate 4c) Kuhl's Pipistrelle (*Pipistrellus kuhli*).

10b. I1 with two points (see figures of teeth a and b under genus *Pipistrellus*, p.149); arm membrane without white margin or only slightly lighter and whitish; ears blackish (plate 4a, 4b)

11a. Forearm 32 to 37 mm; 5th finger over 42 mm (females), 41 mm (males); P1 well developed, visible from outside, I2 longer than short point of I1 (see figure b of *Pipistrellus* teeth, p.149); dorsal side of tail membrane hairy up to middle (plate 5k), ear as in plate 4a; underside hairy only along lower leg Nathusius's Pipistrelle (*Pipistrellus nathusii*).

11b. Forearm 28 to 34.6 mm; 5th finger up to 42 mm (females), 41 mm (males); P1 small, displaced to the inside, barely or not visible from outside, I2 shorter than small point of I1 (see figure a of *Pipistrellus* teeth); lower leg and tail membranes not hairy (plate 5h), ear as in plate 4b
Common Pipistrelle (*Pipistrellus pipistrellus*).

12a. Dorsal fur brown, hair two colored, darker at base, lighter at tip; forearm 39 to 46.4 mm; ear as in plate 2f
Leisler's Bat (*Nyctalus leisleri*).

12b. Dorsal fur rufous, hairs of one color

13a. Forearm 48 to 58 mm, ear plate 2e;
Noctule (*Nyctalus noctula*).

13b. Forearm 63 to 69 mm, ear plate 2d...Greater Noctule (*Nyctalus lasiopterus*).

14a. Very small, triangular ears that do not project beyond top of head (plate 2c); hairs on head short, erect, with distinct boundary with adjoining fur on back; 3rd and 4th fingers at rest curved in joint between 1st and 2nd phalanges (plate 6j); second phalanx of 3rd finger about three times as long as first Schreiber's Bat (*Miniopterus schreibersi*).

14b. Ears always longer than wide, extending beyond top of head (plate 3); tragus lanceolate, more or less tapering to point (plate 3)
Genus *Myotis*

15a. Forearm more than 50 mm long

15b. Forearm less than 50 mm long

16a. Ears longer than 26 mm, front margin not curving backward, tip of ear relatively broad (plate 3a); muzzle long, relatively broad in front; forearm 54 to 67 mm
Greater Mouse-eared Bat (*Myotis myotis*).

16b. Ear less than 26 mm long, front margin relatively straight, ear more narrow, tapering to point (plate 3b); muzzle relatively short and pointed; forearm 52 to 61.5 mm

Lesser Mouse-eared Bat (*Myotis blythi*).

17a. Ear more than 20 mm long (plate 3e), when folded forward projecting nearly by half beyond tip of snout; spur straight, with narrow edging of skin, does not reach half length of margin of tail membrane, last tail vertebra free (plate 5i); forearm 39 to 47 mm

Bechstein's Bat (*Myotis bechsteini*).

17b. Ear less than 20 mm long, when folded forward extending at most 5 mm beyond tip of snout

18a. Spur curved into S-shape, about half as long as margin of tail membrane, free margin of tail membrane wrinkled and covered thickly with short, curved bristles (plate 5f); tragus long, light in color, lanceolate, longer than half ear length (plate 3f)

Natterer's Bat (*Myotis nattereri*).

18b. Spur straight or slightly curved on only one side (plate 5c)

19a. Spur includes about a third of margin of tail membrane, at two-thirds to three-quarters of length of margin of tail membrane is a distinct break, which simulates tip of spur (plate 5d, 5e, 5n); feet with long bristles; hind ear margin without distinct indentation (plate 3e)

19b. Spur includes maximum of half margin of tail membrane, no break in margin of tail membrane at two-thirds to three-quarters of length (plate 5c, b); hind ear margin with distinct indentation (plate 3g, 3h, 3i)

20a. Arm membrane begins before ankle on lower leg (plate 5o); tail membrane above and below from legs to about middle (high spur tip) with dark downy hairs (plate 5n, 5o); dorsal fur gray; tragus curved into slight S-shape, reaches at least half ear length (plate 5m); forearm 38 to 44 mm

Long-fingered Bat (*Myotis capaccinii*).

20b. Arm membrane begins at ankle or middle of sole of foot (plate 5d, 5e); tail membrane lacks conspicuous fur above; tragus straight, does not reach half ear length (plate 3d, 3e); dorsal fur brownish, shiny

21a. Forearm 43 to 49.2 mm; tragus clearly shorter than half ear length (plate 3d); tail membrane with fine whitish hairs on underside along lower leg up to spur (plate 5d); arm membrane begins at ankle

Pond Bat (*Myotis dasycneme*).

21b. Forearm 33 to 42 mm; tragus reaches nearly half ear length (plate 3e); tail membrane not hairy; arm membrane begins at the base of the first toe (plate 5e)

Daubenton's Bat (*Myotis daubentoni*).

22a. Tragus does not reach

indentation on hind ear margin (plate 3i); dorsal fur long, woolly, with reddish tinge; spur without edging of skin (plate 5l)

Geoffroy's Bat (*Myotis emarginatus*).

22b. Tragus extends beyond indentation on hind ear margin (plate 3g, 3h); fur dark gray brown or yellowish brown with golden sheen; spur with narrow edging of skin (plate 5c)

23a. Tip of penis thickened (plate 7a); ears, face, and wing membranes brownish, base of front ear margin and tragus light flesh colored (plate 3g); dorsal fur in adult animals usually light brown with golden sheen; lateral cusp on P3 higher or just as high as P2 (figure page 193, left); P2 not clearly smaller than P1 forearm 33 to 39.2 mm

Brandt's Bat (*Myotis brandti*).

23b. Tip of penis not thickened (plate 7b); ears, face, and wing membranes blackish, base of front ear margin not lightened (plate 3h); dorsal fur predominantly gray brown to blackish brown; lateral cusp on P3 lower than P2, P2 clearly smaller than P1; forearm 31 to 37.7 mm Whiskered Bat (*Myotis mystacinus*).

KEY C: FREE-TAILED BATS

There is only one species in Europe, the European Free-tailed Bat (*Tadarida teniotis*), A photo of the ear is on plate 2i.

PLATE 1:

a , b	Greater Horseshoe Bat
c , d	Lesser Horseshoe Bat
e , f	Mediterranean Horseshoe Bat
g , h	Mehely's Horseshoe Bat
i , j	Blasius's Horseshoe Bat

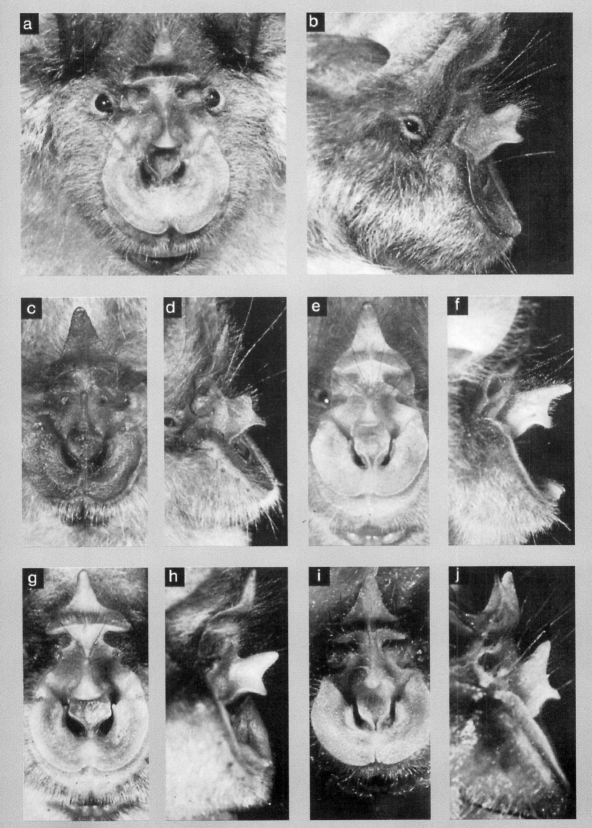

189

PLATE 2:

a   Mediterranean Horseshoe Bat

b   Mehely's Horseshoe Bat

c   Schreiber's Bat

d   Greater Noctule

e   Noctule

f   Leisler's Bat

g   Common Long-eared Bat

h   Gray Long-eared Bat

i   European Free-tailed Bat

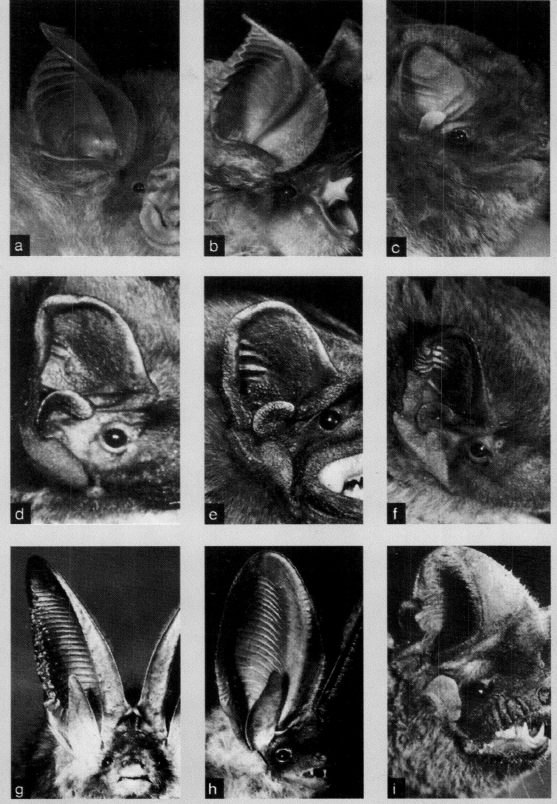

PLATE 3:

EAR FORMS

a	Greater Mouse-eared Bat
b	Lesser Mouse-eared Bat
c	Bechstein's Bat
d	Pond Bat
e	Daubenton's Bat
f	Natterer's Bat
g	Brandt's Bat
h	Whiskered Bat
i	Geoffroy's Bat

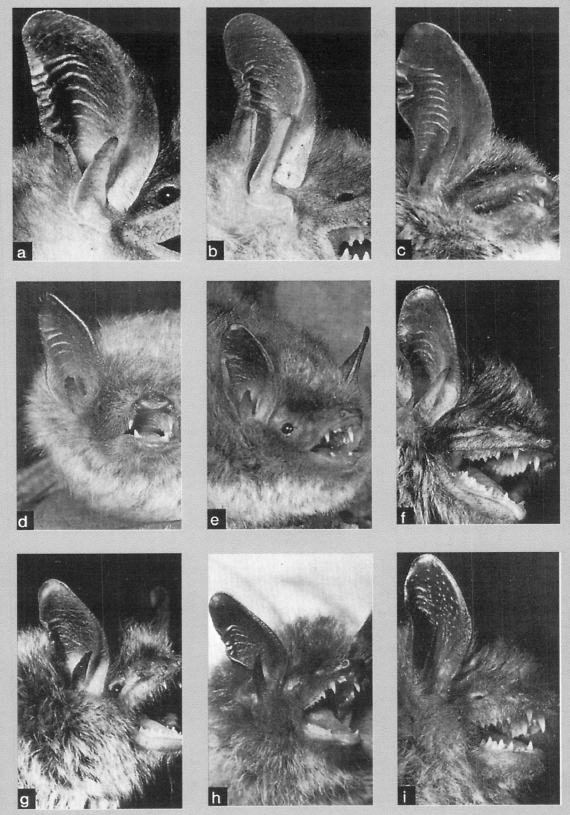

193

PLATE 4:

EAR FORMS

a   **Nathusius's Pipistrelle**

b   **Common Pipistrelle**

c   **Kuhl's Pipistrelle**

d   **Savi's Pipistrelle**

e   **Northern Bat**

f   **Serotine**

g   **Parti-colored Bat**

h   **Barbastelle (with skin fold)**

i   **Barbastelle (without skin fold)**

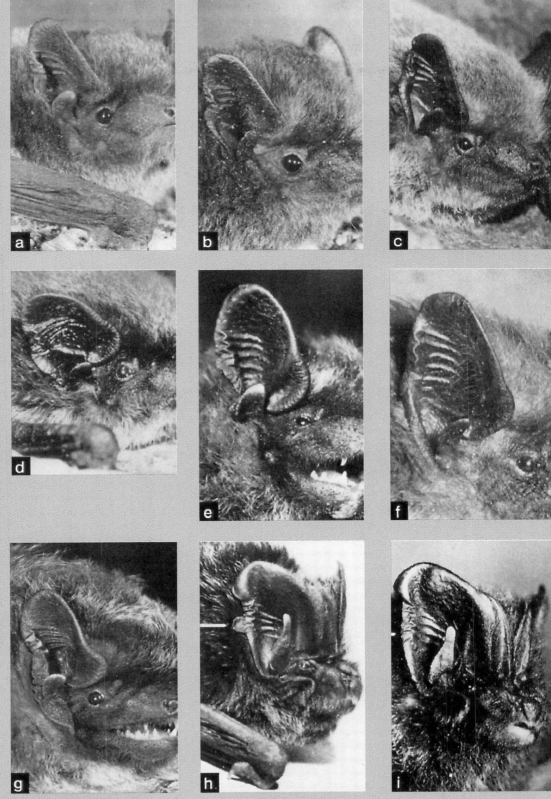

195

PLATE 5:

TAIL MEMBRANES

a   Greater Mouse-eared Bat

b   Lesser Mouse-eared Bat

c   Brandt's Bat

d   Pond Bat

e   Daubenton's Bat

f   Natterer's Bat

g   Common Pipistrelle (underside)

h   Common Pipistrelle (dorsal side, arrow = boundary of fur)

i   Bechstein's Bat

j   Nathusius's Pipistrelle (underside)

k   Nathusius's Pipistrelle (dorsal side, arrow = boundary of fur)

l   Geoffroy's Bat

m   Long-fingered Bat, ear

n   Long-fingered Bat, dorsal

o   Long-fingered Bat, ventral

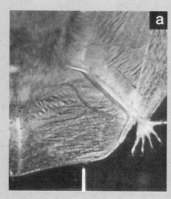
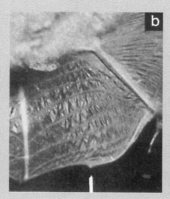
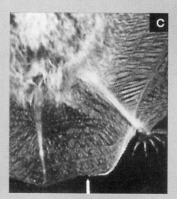

The arrow shows the end of the spur (in d and e the break in the tail membrane).

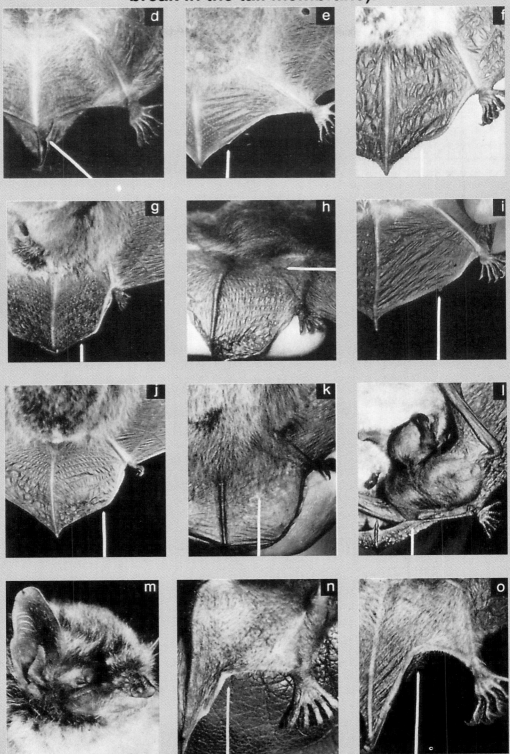

PLATE 6:

TAIL MEMBRANES AND WINGS

a Barbastelle

b Serotine

c Noctule

d Common Long-eared Bat

e Gray Long-eared Bat

f Schreiber's Bat
(The arrows always show the end of the spur.)

g Mediterranean Horseshoe Bat, sleeping posture, third to fifth fingers are bent into the first finger joint, arrow = first joint of the third finger

h Lesser Horseshoe Bat, sleeping posture, arrow = first joint of the third finger

i Natterer's Bat, the third to fifth fingers are bent into the basal joint (still incomplete here), arrow = basal joint of the third finger

j Schreiber's Bat, only the third to fifth fingers are bent into the first finger joint, arrow = first phalanx of the fourth finger

k Mediterranean Horseshoe Bat, bending of the finger, arrow = first phalanx of the fourth finger

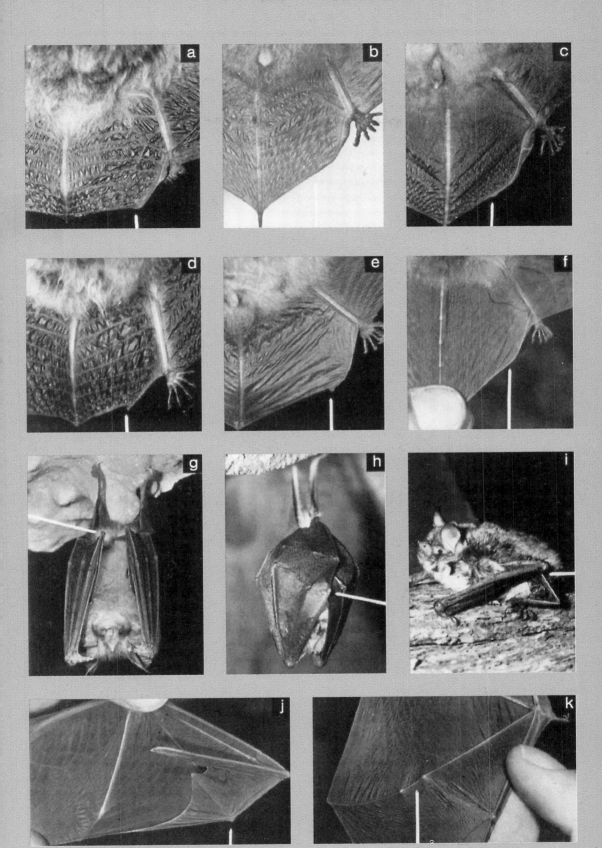

199

PLATE 7:

a   penis, Brandt's Bat

b   penis, Whiskered Bat

c   penis, Common Long-eared Bat

d   penis, Gray Long-eared Bat

e   penis, Nathusius's Pipistrelle; during the mating season the testes and epididymis (arrow) protrude clearly

f   penis, Noctule

g   thumb, Common Long-eared Bat

h   thumb, Gray Long-eared Bat

i   false teats (arrow), Greater Horseshoe Bat, female, right arrow = milk teat, left arrow = false teat

j   teats of a nursing female of Brandt's Bat; to the above left is the beginning of the arm membrane, and to the above right part of the ear is visible

k   nursing, about two-day-old Common Pipistrelle (the mother's wings were raised for presentation)

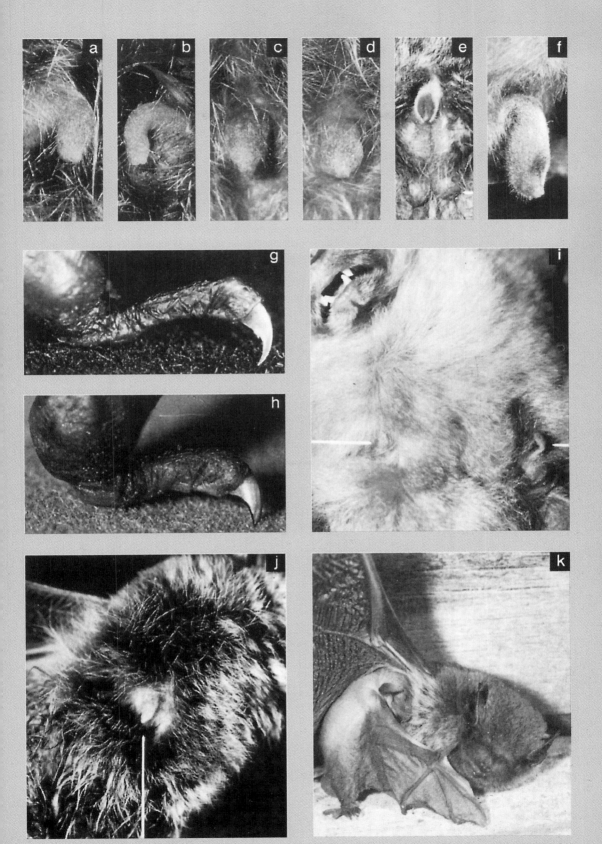

201

SONOGRAMS

FROM HELLER AND WEID

Greater Horseshoe Bat

Mediterranean Horseshoe Bat

Blasius's Horseshoe Bat

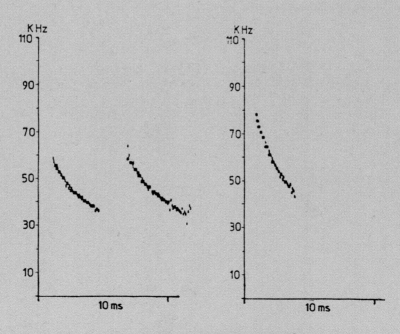

Bechstein's Bat (two different types of calls)

SONOGRAMS

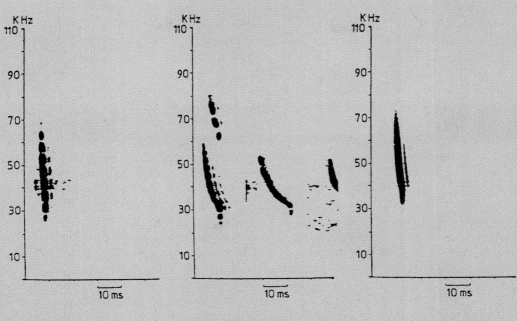

Daubenton's Bat   Pond Bat   Whiskered Bat

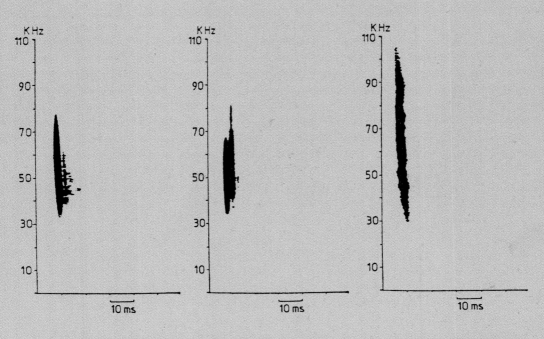

Natterer's Bat   Bechstein's Bat   Mouse-eared Bat

SONOGRAMS

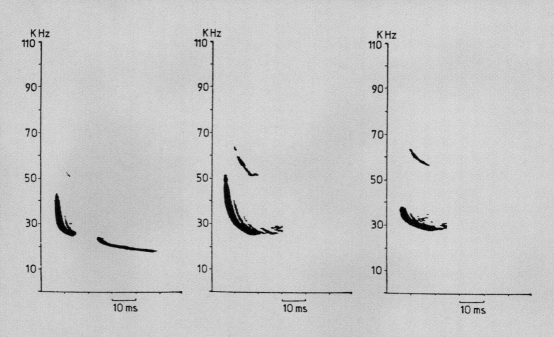

Noctule   **Serotine**   **Northern Bat**

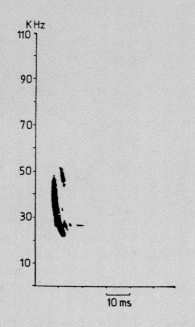

Parti-colored Bat   **Common Long-eared Bat**

SONOGRAMS

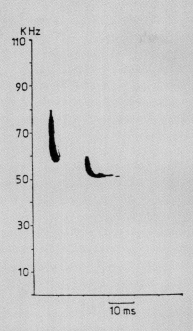

Common Pipistrelle

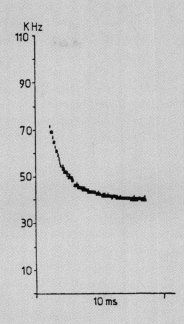

Nathusius's Pipistrelle

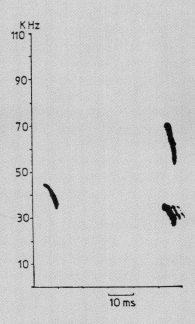

Barbastelle

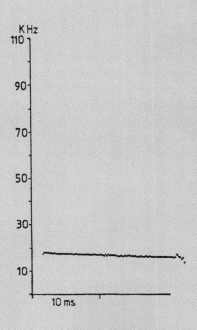

European Free-tailed Bat

TABLE 1:

EUROPEAN BATS AT A GLANCE

Species	maternity roost biotope	migration
Lesser Horseshoe Bat	caves(house)	permanent resident
Greater Horseshoe Bat	caves (house)	permanent resident
Mediterranean Horseshoe Bat	caves (house)	permanent resident
Blasius's Horseshoe Bat	caves	permanent resident
Mehely's Horseshoe Bat	caves	permanent resident
Daubenton's Bat	forest (house)	partial migrant
Long-fingered Bat	caves	?
Pond Bat	house	partial migrant
Brandt's Bat	forest (house)	partial migrant
Whiskered Bat	house	permanent resident(?) (partial migrant?)
Geoffroy's Bat	house (caves)	permanent resident
Natterer's Bat	forest (house)	permanent resident
Bechstein's Bat	forest	permanent resident
Greater Mouse-eared Bat	house (caves)	partial migrant
Lesser Mouse-eared Bat	caves (house)	partial migrant
Noctule	forest	migrant
Leisler's Bat	forest	migrant
Greater Noctule	forest	migrant (?)
Serotine	house	permanent resident (partial migrant?)
Northern Bat	house (?)	permanent resident (?)
Parti-colored Bat	cliffs (house)	migrant
Common Pipistrelle	house (forest)	permanent resident
Nathusius's Pipistrelle	forest	migrant
Kuhl's Pipistrelle	house (cliffs)	permanent resident(?)
Savi's Pipistrelle	house (cliffs?)	partial migrant or migrant
Common Long-eared Bat	forest (house)	permanent resident
Gray Long-eared Bat	house	permanent resident
Barbastelle	forest (house)	partial migrant
Schreiber's Bat	caves	migrant
European Free-tailed Bat	cliffs	partial migrant (migrant?)

temperature in the winter roost	number of young	maximum known age
6° to 9°C	1	21 years
7° to 11°C	1	30 years
approx. 10°C	1	?
?	1	?
?	1	?
(-2°) 3° to 8°C	1	20 years
?	1	?
0.5° to 7.5°C	1	19 years 6 months
(0°) 3° to 7.5°C	1	19 years 8 months
2° to 8°C	1	19 years
6° to 9°C	1	16 years
(-0.5°) 2.8° to 8°C	1	17 years 6 months
3° to 7°C	1	21 years
3° to 12°C	1	22 years
6° to 12°C	1	13 years
0°C to ?	2 (1 to 3)	12 years
?	2	9 years
?	2 (1 to 2)	?
2° to 4°C	1 (1 to 3)	19 years 3 months
1° to 5.5°C	2 (1 to 2)	14 years 6 months
?	2 (2 to 3)	5 years
2° to 6°C	2	16 years 7 months
?	2	7 years
?	2	7 years
?	2	?
2° to 5°C	1 (1 to 2)	22 years
2° to 9°C	1	14 years 7 months
(-3°) 2° to 5°C	2	23 years
7° to 12°C	1	16 years
?	1	more than 10 years

TABLE 2:

Systematics

Suborder Megachiroptera	**Flying foxes**
Family Pteropodidae	flying foxes
Suborder Microchiroptera	**Bats**
Superfamily Emballonuroidea	**Primitive Common Sheath-Tailed Bats**
Family Rhinopomatidae	Mouse-tailed bats, long-tailed bats
Family Craseonycteridae	Hog-nosed bats , butterfly bats, bumblebee bats
Family Emballonuridae	Sheath-tailed bats, sac-winged bats
Family Noctilionidae	Fisherman bats, bulldog bats
Superfamily Rhinolophoidea	**Old World leaf-nosed bats**
Family Nycteridae	Slit-faced bats, hollow-faced bats
Family Megadermatidae	False vampire bats, yellow-winged bats
*Family Rhinolophidae	Horseshoe bats
Family Hipposideridae	Old World leaf-nosed bats, trident bats
Superfamily Phyllostomoidea	**New World leaf-nosed bats**
Family Phyllostomidae	New World leaf-nosed bats, spear-nosed bats
Family Mormoopidae	leaf-chinned bats, naked-backed bats
Superfamily Vespertilionidae	**Higher Common Bats**
*Family Vespertilionidae	common bats, vesper bats, evening bats
Family Natalidae	funnel-eared bats, long-legged bats
Family Furipteridae	thumbless bats, smoky bats
Family Thyropteridae	disk-winged bats, New World sucker-footed bats
Family Myzopodidae	Old World sucker-footed bats
Family Mystacinidae	New Zealand short-tailed bats
*Family Molossidae	free-tailed bats

Footnote: *Families with species that occur in Europe

Species living today	range	diet
175	only the Old World, tropics and subtropics of Africa to Australia	fruits, flowers, pollen
3	northern Africa, Asia to Sumatra	insects
1	Southeast Asia (Thailand)	insects
50	worldwide (pantropic)	insects
2	Central and South America (tropics)	insects, fishes
12	Africa, eastern Asia	insects
5	Africa, Asia, Australia (tropics)	insects, small vertebrates
70	Old World	insects
60	Africa, Asia, Australia (tropics)	insects
148	Central and South America	insects, small vertebrates, fruits, pollen, blood
8	Central and South America	insects
320	Worldwide (Nearctic, Palearctic, southern Africa, Australia, Hawaiian Islands, New Zealand)	insects, individual species: fishes, small vertebrates
6	Central America, Caribbean, insects	
2	Central America, tropical South America	insects
2	Central America, tropical South America	insects
1	Madagascar	insects
1	New Zealand	insects
90	worldwide (primarily pantropic, also southern Europe and southern North America	insects

NORTH AMERICAN BATS

Bats are mammals that fly by use of their "hand wings" (Chiroptera). All living bat species belong to one of the two major suborders, Megachiroptera (flying foxes of the Old World tropics) or Microchiroptera (bats). The latter are found worldwide. There are approximately 980 different species of bats and flying foxes. The majority of them inhabit tropical forests.

Forty-four species belonging to four families live in North America. The evening bats (plain-nosed bats) belong to the family Vespertilionidae, the free-tailed bats belong to the family Molossidae, the leaf-nosed bats belong to the Phyllostomatidae, and the leaf-chinned bats belong to the Mormoopidae.

The largest family in North America, the Vespertilionidae (evening bats), includes some 30 species. Not all of the species are distributed regularly throughout the range. Some of them are common and a few are quite rare. There are more species in the warm southern states than in the northern states as not all species are well adapted (by migration and/or hibernation) to the cold winters.

Bats are found in an amazing variety of sizes. Their appearance is more fascinating than strange. North American bats also present very diverse portraits, including some with enormous ears or nose leaves.

About 70% of bats feed on insects. Worldwide they are the major predators of night-flying insects (including numerous pests). Many tropical species feed on fruits and/or nectar, and a few are carnivorous.

More than half of the bat species in North America are endangered.

Many bat species have been poorly studied because they are difficult to find so that the habits of even relatively common bats are not well known. Therefore, in the following pages only 39 of the 44 North American species are presented, accompanied only by a very short description.

Abbreviations:

HB = Head-body length
T = Tail length
H = Hind foot length
FA = Forearm length
Wt = Weight

EVENING OR PLAIN-NOSED BATS, FAMILY VESPERTILIONIDAE

MYOTIS LUCIFUGUS
Little Brown Bat

The fur of the dorsal side is glossy brown to russet to gray, the ventral side is buff. The tragus short and rounded and the calcar has or does not have a narrow keel. This is one of North

America's most abundant and widespread species.

Measurements:
HB 79-93 mm
T   31-40 mm
HF 8.5-10 mm
FA 30-45 mm
Ear 14-16 mm
Wt 3.1-14.4 g

Habitat: In the summer this bat roosts (has nurseries) in buildings, as well as caves and mines. In the winter they have hibernacula in caves and mines, hibernating in irregular clusters. This is a migratory species, with movements of several hundred miles between the summer and winter roosts.

Range: The distribution of this species includes much of North America. It extends north to mid-Alaska, and south to most states except Florida, Texas, and southern California.

Reproduction: There is only one young per female. The nursery colonies often include 300-800 (or more) bats. This species is very heat tolerant.

Hunting and feeding: the diet consists primarily of small insects (e.g. flies, moths). that they often catch above rivers or lakes.

MYOTIS YUMANENSIS
Yuma Bat

The fur is short and dull colored, the dorsal region being shades of brown, the ventral area lighter. The calcar does not have a keel.

Measurements:
HB 84-99 mm
T   32-45 mm
HF 9-11 mm
FA 33-37 mm
Wt 5-7 g

Habitat: This bat inhabits moist areas (streams, lakes, ponds). It roosts in buildings or under shingles, and the nursery colonies are in roofs of buildings, caves, mines, and under bridges.

Range: The Yuma Bat is distributed in western North America. It extends east to western Montana, south to Colorado, and into New Mexico.

Hunting and feeding: It hunts close to the surface of the water. its diet includes small insects (ex. moths, termites).

MYOTIS AUSTRORIPARIUS
Southeastern Bat

The fur is short, thick, and woolly. It is a dull brown, the base of the hair is buff colored, and the belly is often white (in Indiana there is a white populations). The calcar is not keeled.

Measurements:
HB 84-99 mm
T   36-45 mm
HF 10-12 mm
FA 35-42 mm
Wt 7-12 g

Habitat: In the north it prefers caves, in the south it is found in buildings as well as hollow trees. In Florida caves, too, are used during the wintertime and they are often found in culverts, beams of bridges, and buildings.

Range: This species occurs in the southeastern United States, as well as southern Illinois and southern Indiana.

Reproduction: Nursery

Myotis grisescens, the Gray Bat

colonies are formed in March. Caves in Florida may include colonies of up to 90,000 individuals.

MYOTIS GRISESCENS
Gray Bat

Medium sized. The fur is unicolored grayish or brownish. The calcar is without a keel. The ears are short; the tragus is relatively short and rounded.

Measurements:
HB 80-96 mm
T   32-44 mm
HF 8-11 mm
FA 41-46(?) mm
Wt 6-9 g

Habitat: This is truly a cave species preferring caves in moist areas. They migrate between summer and winter caves in large groups, and may travel as far as 300 miles.

Range: In the south they occur in Alabama, western Georgia, Tennessee, and Kentucky while in the north the range extends to southern Indiana, southern Illinois, and Missouri.

Reproduction: Females and young hang in clusters separate from the males. They disband in late July.

Hunting and Feeding: The diet consists entirely of insects.

MYOTIS VELIFER
Cave Bat

Large species. In the eastern part of its range the fur is light brown, but rather black in the western part. The ears reach the tip of the nose when extended forward. The calcar is not keeled.

Measurements:
HB 90-115 mm
T   41-49 mm
HF 10-12 mm
FA 40-47 mm
Ear 15-17 mm

Habitat: In the summer this species inhabits caves, mines, and sometimes buildings; in the winter it is found in caves, in tight clusters on walls or ceilings, seldom in crevices. The Cave Bat occasionally migrates between summer and winter roosts.

Range: The distribution includes the arid southwest: Texas, Oklahoma, southern Kansas, New Mexico, and Arizona.

Reproduction: there is a large nursery roost in Kansas with 15,000-20,000 animals.

MYOTIS KEENII
Keen's Bat

The fur is brown to pale brown, and the belly is yellowish. The ears are long, extending beyond the nose about 4 mm when folded forward. The tragus is thin and long; the calcar is not keeled.

Measurements:
HB 79-88 mm
T   36-45 mm
HF 7-9 mm
FA 34-39 mm

Ear 17-19 mm
Wt 5-10 g

Habitat: In the east the summer roosts are under loose bark or behind shutters. In winter it roosts in caves and mines, singly or in small clusters. In the west they are found in dense forests.

Range: This species occurs on the west coast of Washington and British Columbia. In the east it is found from Newfoundland to Georgia. In the central states it extends from Saskatchewan to Arkansas.

Reproduction: Maternity roosts have about 30 females located behind bark.

Hunting and Feeding: This species feeds on small insects (such as flies).

MYOTIS EVOTIS
Long-eared Bat

The fur is long, glossy, and brown. The ears are dark (black), longer than in any other species of *Myotis*. They extend beyond the nose about 7 mm when folded forward.

Measurements:
HB 75-97 mm
T   36-45 mm
HF 7-10 mm
FA 35-41 mm
Ear 22-25 mm

Habitat: This species prefers coniferous forests of high mountains, and is seldom found in buildings. They occur singly or in small groups behind bark (by day); at night they also roost in caves.

Range: The range includes western North America (from

southern British Columbia to western New Mexico).

Hunting and Feeding: The prey consists primarily of small moths, flies, and beetles.

MYOTIS AURICULUS

The fur is dull brownish. The ears are large, brown in color. The tragus is long and thin. The calcar is not keeled.

Measurements:
HB 78-88 mm
T   36-45 mm
HF 7-9 mm
FA 38-40 mm
Ear 14-20 mm

Habitat: Deserts, chaparral to forests; areas of rocky outcroppings.

Range: southern Arizona and southern New Mexico.

MYOTIS THYSANODES
Fringed Bat

The fur on the dorsal side is brown to reddish brown, on the ventral side it is lighter. The free edge of the tail (interfemoral) membrane is covered with bristles (a fringe of hair). The ears are long.

Measurements:
HB 80-95 mm
T   37-42 mm
HF 8-11 mm
FA 36-46(?) mm
Ear 16-20 mm

Habitat: This species prefers protected locations, such as desert scrub, oak, pinyon, and juniper forests. It roosts in caves, mines, and buildings. Its winter habits are unknown.

Range: The distribution is the southwestern United States (Texas, Arizona, Colorado, Utah to California).

Reproduction: Nursery colonies in the low hundreds exist; the males are absent.

Hunting and feeding: The diet is mainly moths, crickets, and harvestmen.

MYOTIS SODALIS
Indiana Bat or Social Bat

The fur is pinkish brown, and the lips and nose are pinkish. The tragus is short and rounded. The calcar has a keel.

Measurements:
HB 71-79 mm
T   27-44 mm
HF 7-9 mm
FA 35-41 mm
Wt 5-8 g

Habitat: In the summer it can be found in wooded or semi-wooded areas along streams. In the winter it inhabits caves, where they hibernate from October in tightly packed clusters. In spring they migrate to the summer roosts.

Range: The range is from the midwest to the east, and extending in the south to Alabama.

Hunting and feeding: The diet consists of small insects.

MYOTIS VOLANS
Long-legged Bat

Large species. The fur on the dorsal side is tawny or reddish to black, the ventral side is grayish to pale buff. In the west the calcar has a well-developed keel, the forearm is hairy to the elbow, and the tail membrane is hairy to the knee. The ears are short.

Measurements:
HB 87-103 mm
T   37-49 mm
HF 8-10.5 mm
FA 35-42 mm

Habitat: In the summer it occurs in trees, crevices, and buildings, but particularly in forest areas. The behavior in the winter is unknown.

Range: This species is distributed in the western half of North America. In the north it extends to British Columbia.

Reproduction: There are nursery colonies with up to several hundred animals.

Hunting and feeding: The diet consists of small insects (moths).

MYOTIS CALIFORNICUS
California Bat

Fur: dull, light to dark brown; dorsal side yellowish to orange cast; ventral side paler. Ears, wings, tail membrane dark; feet tiny; calcar keeled.

Measurements:
HB 74-85 mm
T   36-42 mm
HF 5-7 mm

Habitat: Desert to semi-desert areas; rocky canyons in southwest; during daytime: buildings, bridges, hollow trees, under bark; during nighttime: buildings; winter: hibernate in mines, some remain active.

Range: Western third of North America; in the north to British Columbia, in the south to New Mexico.

Hunting and feeding: Conspicuously, erratic flight; food: small flies, moths.

MYOTIS LEIBII
Small-footed Bat

Fur: glossy, dorsal side golden brown, ventral side buff to white; wings and tail membrane dark brown; calcar keeled; ears black, face black, hind foot slightly smaller than in other species of *Myotis*.

Measurements:
HB 71-82 mm
T   30-38 mm
HF 6-8 mm
FA 30-36 mm
Wt 6-9 g

Habitat: Winter in caves on floor or wedged into crevices, also below rocks.

Range: Western half of United States, in the east isolated populations.

LASIONYCTERUS NOCTIVAGANS
Silver-haired Bat

Medium sized species. The fur is dark, either brown or black, with silvery tipped hairs on the dorsal side, giving it a "frosted" appearance. The tail membrane is covered with fur only on the anterior half. The ear is short, rounded.

Measurements:
HB 92-198 mm
T   37-45 mm
HF 9-10 mm
FA 37-44 mm

Habitat: In the summer it can be found in forested areas (behind bark, in tree caves); in the winter it hibernates in tree hollows, woodpiles, or crevices. It also migrates from north to south between the summer and winter roosts for distances of more than 100 miles.

Lasionycterus noctivagans.

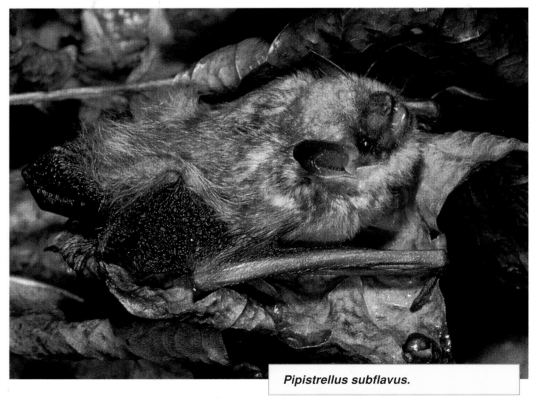

Pipistrellus subflavus.

Eptesicus fuscus.

Range: This species occurs in southern Canada and most of the United States, exclusive of the southern regions.

Reproduction: Twins are usually born in late July.

Hunting and feeding: The bats emerge early in the evening. They fly very slowly, catching a wide variety of small insects.

PIPISTRELLUS HESPERUS
Western Pipistrelle

This is the smallest bat in the United States. The fur on the dorsal side is light yellow, or

Lasiurus borealis.

grayish to reddish brown. On the ventral side it is whitish. This is in clear contrast to the black face mask, and the blackish wings, tail membrane, ears, and feet. The tagus is rounded, the calcar keeled.

Measurements:
HB   60-86 mm
T   27-32 mm
HF   6-7 mm
FA   27-33 mm
Wt   5-6 g

Habitat: In the summer it inhabits buildings, caves, rocky areas, and scrub. It hibernates in caves, mines, and crevices. Some individuals do migrate.

Range: This species is found in the southwestern United States and in the north to southeastern Washington.

Hunting and feeding: This is the first bat to appear in the evening (often it even flies in the daylight). It has a slow and jerky flight. Its food consists of tiny insects.

PIPISTRELLUS SUBFLAVUS
Eastern Pipistrelle

This is the smallest species in the east. Its fur is reddish to light brown. The hairs tricolored: at the base almost black, in the middle there is a band of light yellowish brown, and at the tip there is a darker contrasting color The tragus is blunt, and the calcar does not have a keel.

Measurements:
HB   81-89 mm
T   36-45 mm
FA   31-35 mm
Wt   3.5-6 g

Habitat: The bats hang in vegetation during the daytime. Nursery roosts are also in buildings. In the fall they migrate to winter roosts. They hibernate in caves and mines, and in crevices.

Range: This species range includes the central and eastern part of the United States.

Hunting and feeding: Their diet consists of tiny insects.

EPTESICUS FUSCUS
Big Brown Bat

Large species. It is very common, being widely distributed in the United States and Canada. The fur on the dorsal side is a glossy brown (from light to dark), the ventral side is paler. The wings and tail membranes are black. The calcar is keeled.

Measurements:
HB   106-127 mm
T   42-52 mm
HF   10-11.5 mm
FA   42-51 mm
Wt   13-18 g

Habitat: In the summer it resides in buildings, or sometime in tree hollows. It is one of the most urban species of bats in North America. In the winter it lives in caves, mines, and buildings. They hang singly or in groups near the entrance. This bat is extremely hardy, even hibernating as far north as Canada. They can travel up to 150 miles.

Range: Their distribution covers southern Canada and the entire United States, with the exception of Florida and southern Texas.

Reproduction: Nursery colonies include 20-300 individuals. One or two young are born in late May or early June.

Hunting and feeding: This is a high flying bat with plenty of speed. The diet includes a wide variety of insects (large beetles, wasps, ants, planthoppers, flies, moths).

LASIURUS BOREALIS
Red Bat

This is a solitary bat. The fur is long, angora-like. Males are a bright orange to yellowish brown; females are dull red, brick red, or chestnut. Both sexes have a frosted appearance on the backside and breast. White markings can be seen on the shoulders and wrists. The ears are small and rounded, the wings long and narrow. The upper surface of the tail membrane is densely covered with fur. This is one of the few bats with four teats.

Measurements:
HB 95-126 mm
T 45-62 mm
HF 8.5-10 mm
FA 36-44 mm
Wt 9.5-15 g

Habitat: In the summer this bat lives amid the dense foliage of deciduous trees of hedgerows or at forest edges. They hang by one foot 4-10 inches above the ground. Females with young roost 10-20 inches above the ground. They are well adapted to surviving at low temperatures; they wrap themselves in their large furry tail membrane.

Northern populations migrate to the southern part of the range.

Range: They range from Canada and the United States throughout most of Latin America.

Reproduction: Mating takes place in flight. Often triplets and even quadruplets are born.

Hunting and feeding: These are fast flyers that start hunting in the early evening. Their food consists of different kinds of insects (moths, beetles, planthoppers, ants, flies).

LASIURUS SEMINOLUS
Seminole Bat

The fur on the dorsal side is brown (deep mahogany) with silver white tips. The head is reddish brown. The surface of the tail membrane is covered with fur. The ears are short and round.

Measurements:
HB 108-114 mm
T 44-52 mm
HF 8-9 mm
FA 35-45 mm
Wt 7-14 g

Habitat: During the daytime the animals hang 3.5-5 inches above the ground in clumps of Spanish Moss.

Range: They occur in the southeastern United States and the southern Atlantic Coast.

Reproduction: Three to four young are born late May to early June.

LASIURUS CINEREUS
Hoary Bat

This is the largest, most handsome species. It is a solitary

Lasiurus cinereus.

Plecotus townsendii.

Euderma maculatum.

Tadarida brasiliensis.

Eumops perotis.

bat. The fur of the dorsal side is light brown with white tips, and the throat is yellowish. The ears are short and rounded, with naked black rims. The dorsal surface of the tail membrane is covered with dense fur. It has unusual fur-lined wings, and two pairs of teats.

Measurements:
HB 102-152 mm
T   44-65 mm
HF 6-14 mm
FA 44-59 mm
Wt 20-35 g

Habitat: They hang on evergreen branches, roosting in foliage. The northern populations perform long seasonal migrations. In cold climates they also hibernate in caves, in the southern U. S. they hibernate on tree trunks.

Range: The range includes southern Canada and the entire United States, except Florida. This is the most widely distributed bat in the United States, but it is not as common in the eastern states. This is the only bat of Hawaii.

Reproduction: They mate in fall (!) while flying. Twins are born naked and blind in May/June. Sometimes there are 3 or 4 young.

Hunting and feeding: They emerge in the late evening. They feed on moths.

LASIURUS INTERMEDIUS
Northern Yellow Bat

This is a large, solitary species. The long silky fur is yellowish brown with black hair tips on top. The tail membrane is covered with fur on the basal half of the dorsal side. The ears are broad and rounded. The tragus is broad based, with a tapered tip. The total length (tip of nose to tip of tail) is more than 120 mm.

Measurements:
HB 115 mm
T   51 mm
HF 10.25 mm
FA 45-56 mm

Habitat: This tree loving species roosts singly by day in clumps of Spanish Moss in areas of pine and oak.

Range: It inhabits the southeastern United States, i.e., the coastal region from North Carolina to eastern Texas.

Reproduction: Females form loose colonies during breeding. The normal litter size is three young.

Hunting and feeding: This bat forages 15-25 inches above ground along forests and open grassy areas.

LASIURUS EGA
Southern Yellow Bat

This is a large species, but it is smaller than the Northern Yellow Bat. Its total length is less than 120 mm. The fur is yellowish or orange to brownish or gray. The anterior half of the upper surface of the tail membrane is covered with fur. The ears are rather large, outside partially covered with fur.

Measurements:
HB—
T   —
HF 10 mm

FA 45-48 mm
Ear 17 mm
Wt 9.2-22.5 g
Habitat: This species roosts in various trees.

Range: It is found in extreme southern California and in southern Arizona.

Hunting and feeding: It emerges early in the evening from its day roost.

NYCTICEIUS HUMERALIS
Evening Bat

Description: The fur of the dorsal side is reddish brown to dark brown; the ventral side is tawny. The calcar does not have a keel. The tragus is short and rounded. There is only one upper incisor.

Measurements:
HB 78-93 mm
T   35-37 mm
HF 7-10 mm
FA 33-39 mm
Wt 5-9 g

Habitat: This species is found in the summer in buildings, hollow trees, and in Spanish Moss. The winter habitat is unknown. Some move to the south during the fall. This is one of the most abundant bats around towns in the southern coastal states.

Range: The distribution includes the central and southeastern United States.

Reproduction: Maternity colonies sometimes include several hundred individuals. There is a single young.

Hunting and feeding: This bat flies low at night, feeding on small insects.

EUDERMA MACULATUM
Spotted Bat (also called Death's Head Bat because of its striking coloration)

Description: This is a beautiful species. The fur of the dorsal side is black (dark brown) and there are three large white spots (one on each shoulder and the third at the base of the tail). The ventral side is white. The strikingly long ears are pink. They are curled and lie backward when the bat rests. This is a rare bat in North America.

Measurements:
HB 107-115 mm
T   46-50 mm
FA 48-51 mm
Ear 51 mm

Habitat: They live in crevices of rocky cliffs and canyons. They hibernate in small clusters.

Range: This species is found in Arizona, New Mexico, and southern California.

Hunting and feeding: It emerges late in the evening, feeding on moths.

PLECOTUS TOWNSENDI
Townsend's Big-eared Bat

The fur of the dorsal side is pale gray or brown, very dark at the base. The ventral side is buff. It has enormous large ears that extend to the middle of the body when laid back. There are two large glandular lumps behind the nostrils.

Measurements:
HB 89-110 mm
T   35-54 mm
HF 10-13 mm
FA 39-47 mm
Ear 31-37 mm
Wt 9-12 g

Macrotus californicus.

Laptonycteris curasoae.

Habitat: This bat lives in caves and buildings. In the west it inhabits scrub deserts, pine and pinyon/juniper forests; in the east it lives in oak/hickory forests. In the summer nursery colonies with 200-1000 females are seen; in the winter it lives in caves. During hibernation their big ears are folded onto the back.

Range: The range includes the western third of United States.

Hunting and feeding: This bat emerges late in the evening. Its food is moths.

PLECOTUS RAFINESQUII
Rafinesque's Big-eared Bat

The fur is brown, darkest at the base, the underside has white tips. The ears are very large. When resting, they coil their ears against the side of the head like a ram's horn. There are two large glandular lumps on the nose.

Measurements:
HB 92-106 mm
T   41-54 mm
HF 8-12 mm
FA 40-46 mm
Ear 32-36 mm
Wt 9-12 g

Habitat: They inhabit buildings in forest regions. In the north they prefer caves or mines. They roost singly.

Range: The range includes the southeastern United States, in the west to Louisiana and in the north to Kentucky.

Hunting and feeding: Their hovering flight enables them to pluck insects from foliage.

IDIONYCTERIS PHYLLOTIS
Allen's Big-eared Bat

The fur is tawny on the dorsal side, the hair having a dark brown base; the ventral side is lighter. There are white patches behind the ears. The ears are enormously long; when resting the ears are folded back. There are two skin flaps on the base of the inner ear margin (only in this big-eared bat).

Measurements:
HB 103-118 mm
T   46-55 mm
HF 10-11 mm
Ear 40 mm
Wt 10.4-13.2 g

Habitat: They are found in caves and mines in wooded mountains.

Range: The range includes Arizona and western New Mexico.

Hunting and feeding: It emerges late in the evening.

ANTROZOUS PALLIDUS
Pallid Bat

Large species. The dorsal fur is yellowish, cream to beige, the fur being lightest at its base. The ventral side is white. All other big-eared bats are much darker. The ears are big, separated at the base.

Measurements:
HB 107-130 mm
T   35-39 mm
HF 11-16 mm
FA 48-60 mm
Ear 25-36 mm
Wt 28-37 g

Habitat: Arid regions are preferred. In the daytime they can be found in buildings, bridges, deep crevices in rock faces, but seldom in caves, mines, and hollow trees. In the summer colonies of 30-100 animals include adults of both sexes as well as young ones. There are no lengthy migrations.

Range: It occurs from Washington to southern California, as well as southern Kansas and western Texas.

Reproduction: Twins are normally born in June.

Hunting and feeding: This bat emerges late at night. It primarily feeds on the ground. The diet includes large insects (flightless beetles, crickets, scorpions, grasshoppers), lizards (?), and nectar (?).

FREE-TAILED BATS, FAMILY MOLLOSIDAE

TADARIDA BRASILIENSIS
Mexican (Brazilian) Free-tailed Bat

This is the smallest free-tailed bat. The fur of the dorsal side is dark brown or dark gray, the hair whitish at the base. The lips are vertically wrinkled along the muzzle. The ears are medium sized and broad, and not joined at their bases. At least a third of the tail extends freely from the tail membrane. The backward pointed calcar reaches more than half the length of the short tail membrane.

Measurements:
HB 90-110 mm
T   33-44 mm
FA 36-46 mm
Wt 11-14 g
Habitat: This species lives in

buildings in the southeast and on the west coast. It is found in caves, buildings, and bridges from Texas to Arizona. This is the most common bat in the southwest, and one of the most numerous mammals in the country (at least 100 million animals). It forms the largest colonies of any warm-blooded animals. In some caves there are huge populations with 400 bats per square foot (Braken Cave: about 20 million individuals). Only a few hibernate. In the fall most of them migrate south to Mexico (sometimes as much as 1000 miles), where they spend the winter, remaining active.

Range: The range includes the southern United States, extending from southern Oregon in the west to South Carolina in the east.

Reproduction: A single young is born in June. It hangs by itself in the nursery. Mothers nurse their baby after returning from a hunt. The first flights of the young occur after about 5 weeks. The lifespan is 13-18 years.

Hunting and feeding: The evening flight begins 1-2 hours before sunset in great colonies. The food includes mostly moths.

TADARIDA FEMOROSACCA
Pocketed Free-tailed Bat

The fur of the dorsal side is dark gray or brown. The lips along the muzzle are wrinkled. The wings are long and narrow. About half the tail is without a membrane. The ears join at their base. There is an inconspicuous femoral pocket. There are long stiff hairs on their feet.

Measurements:
HB 98-118 mm
T   30-42 mm
FA 44-51 mm

Habitat: This bat inhabits rocky desert areas. By day they are found in rock crevices. Groups usually include less than 100 animals.

Range: This bat is distributed in southern California and southern Arizona.

Hunting and feeding: The diet consists of various insects (moths, ants, wasps, leafhoppers).

NYCTINOMOPS MACROTIS
(*TADARIDA MACROTIS*)
Big Free-tailed Bat

Large species. The fur of the dorsal side is reddish brown, dark brown, or black, the hair bases white. The lips are wrinkled along the muzzle. The tail is free for one inch or more. The ears have their inner edges joined at the base, and they extend beyond the tip of the nose.

Measurements:
HB 129 mm
T   43-144 mm
FA 58-64

Habitat: They prefer rocky areas. By day they live in rocky cliffs.

Range: This species is widespread but rare (uncommon).

Hunting and feeding: It emerges late in the evening. The food consists mainly of insects (moths, crickets, grasshoppers, ants).

EUMOPS PEROTIS
Western Mastiff Bat (also known as Bonnet Bat)

This is the largest bat of North America. It is sparsely covered with fur, and has dark brown hairs, white at the base. The lips are not vertically wrinkled. The ears are large, joined at their base, and extend beyond the nose when folded forward. The wings are long. They often crawl on the ground to search for food, at which time the tail perhaps serves as a tactile organ.

Measurements:
HB 140-185 mm
T   35-80 mm
FA 72-82 mm
Ear 25-40 mm

Habitat: By day they are found in rocky cliffs, in long vertical slits, and in buildings. They roost in crevices and have large droppings. Small colonies may be seen with fewer than 100 animals.

Range: They are found in southern California and southern Nevada.

Hunting and feeding: They forage high, at great distances from their roosting places. Their food is mostly moths, but also ground-living crickets and grasshoppers.

EUMOPS UNDERWOODI
Underwood's Mastiff Bat

Large species. The fur on the dorsal side is dark brown; on the ventral side it is grayish. The lips are smooth along the muzzle. The wings and tail membranes are grayish. The ears are large, with the inner edges joined at the base.

Measurements:
HB 160-167 mm
T   52-60 mm
HF 15-20
FA 67-70 mm
Ear 29-33 mm
Wt 53-61 g

Habitat: This species lives in deserts.

Range: It is found in Arizona and south to northern Central America.

EUMOPS GLAUCINUS
Wagner's Bat

The fur is gray to black, slightly lighter below. The lips are smooth along the muzzle. The free tail extends far beyond the tail membrane. The ears are joined at the base. The wings are long and narrow.

Measurements:
HB 80 mm
FA 57-66 mm

Habitat: It is found under "Cuban tile."

Range: This rare bat is found only in southeastern Florida.

LEAF-NOSED BATS, FAMILY PHYLLOSTOMATIDAE

MACROTUS CALIFORNICUS
California leaf-nosed Bat

The fur on the dorsal side is grayish to dark brownish, paler below. The ears are large (more than 25 mm). There is an erect triangular nose leaf.

Measurements:
HB 84-93 mm

T   more than 25 mm
FA 47-55 mm

Habitat: This bat prefers desert scrub. It roosts year round near the entrances to mines or tunnels in small groups (up to 100 bats), each animal separated from the other. They often hang by one leg from a mine or tunnel ceiling, There is no hibernation.

Range: The range includes southern California, southern Nevada, and southern Arizona.

Hunting and feeding: Various insects (including flightless species) are eaten. They catch their insect prey from the foliage or pluck it from the ground.

CHOERONYCTERIS MEXICANA
Long-tongued Bat

The fur of the dorsal side is gray or brownish, paler below. The eyes are large. The long slender nose has an erect arrowhead-shaped flap of skin. The small, tiny tail extends less than half way to the end of the interfemoral membrane. The tongue is long, and has on the tip a brush of tiny nipple-like processes. They lack lower incisors.

Measurements:
HB 55-78 mm
FA 43-44 mm

Habitat: They prefer canyons in montane regions. By day they are found in caves and mines, seldom in buildings, hanging near the entrance.

Range: They occur in southwestern California, southeastern Arizona, and southwestern New Mexico.

Hunting and feeding: They feed on insects, pollen, nectar, and the juices of fruits.

LEPTONYCTERIS NIVALIS
Mexican Long-nosed Bat

The fur is grayish brown above, paler on the shoulders and on the ventral side. The ears are small. A tail is absent. The long nose has a triangular shaped leaf. The elongated nose and long flexible tongue facilitate the feeding on flowers.

Measurements:
HB 76-88 mm
T   more than 25 mm
FA 55-60 mm
Ear 15 mm
Wt 21 g

Habitat: A pine/oak habitat and caves are preferred.

Range: In the United States they are found only at Emory Peak (Texas).

Hunting and feeding: This bat emerges late in the evening. The food is mostly nectar and pollen.

LEPTONYCTERIS SANBORNI
Sanborn's Long-nosed Bat

The fur of the dorsal side is reddish brown. The tail is not visible. The nose is long and has an erected triangular-shaped skin lobe on the tip The eyes are large. The long snout and the long tongue with hair-like processes on the top are ideally suited for insertion into flowers.

Measurements:
HB 69-84 mm
T   more than 25 mm
HF 13-17 mm
FA 51-56 mm

Habitat: They are found in mountains in the desert, in caves and mines. The bats migrate to Mexico in the fall.

Range: They occur in southern Arizona.

Reproduction: Females congregate in large maternity colonies.

Hunting and feeding: This bat emerges late in the evening. The diet includes insects, nectar, and pollen.

LEAF-CHINNED BATS, FAMILY MORMOOPIDAE

MORMOOPS MEGALOPHYLLA
Ghost-faced Bat

The fur of the dorsal side is reddish brown to dark brown. There is no nose leaf. The tail is short. There are folds of skin across the chin from ear to ear. This is unique in North America.

Measurements:
HB 59-66 mm
FA 46-56 mm

Habitat: This bat is found in desert scrub, in caves or mines, seldom in buildings, often in very hot places with a high humidity. They roost in loose colonies, with a colony size of up to 500,000 animals. They do not hibernate.

Range: They occur in southern Texas and southern Arizona.

Hunting and feeding: These are strong flyers. They emerge late in the evening, hunting above the ground.

BATS AND RABIES

There have been many isolated incidents of bats testing positive for rabies. Thanks to Drs. Michelle Feller, John Kaneene and Mary Grace Stobierski (JAVMA Vol. 210 (2), January 15, 1997, p.195), a study was completed on the *Prevalance of Rabies in Bats in Michigan, 1981-1993.* The objective of this study was to analyze the species distribution of animals submitted to the Michigan Department of Public Health (MDPH) for rabies testing during 1993. To determine whether any of the 9 species of bats found in Michigan carries a disproportionate amount of rabies, and to determine whether bats contributed the most cases of confirmed rabies during 1981 through 1992.

During 1993 the MDPH tested 2,045 animals. Seventeen rabid animals were identified. They included one cat, one skunk and 15 bats! There were 246 bats submitted for testing. *Eptesicus fuscus,* the big brown bat, accounted for 239 (97.2%) of the bats tested and the only rabid bats were this species. Annual percentages of submitted bats found to be rabid ranged from 2 to 11% with an average of 6.2% over the 13-year study period.

The conclusion was that 100% of the confirmed cases of rabies in bats reported in Michigan in 1993 were associated with the big brown bat *Eptesicus fuscus.* During the period from 1981 through 1992, most of Michigan's confirmed cases of rabies in animals developed in bats!

RABIES IS A USUALLY FATAL DISEASE TRANSMITTED TO HUMANS BY ANI-MALS. It seems that bats are the most frequent carriers of the disease.

Prepared by Prof. Dr. Herbert R. Axelrod

PHOTO CREDITS

The authors and publisher are most appreciative and thankful to the following photographers for their assistance in illustrating this book. The numbers in parentheses following the photographers' names indicate the number of photos used.

COLOR PHOTOGRAPHS
A. Benk (2)
J. Cerveny (2)
J. Gebhard (2)
E. Grimmberger (79)
K.G.Heller (2)
O.v. Helversen (2)
A. Limbrunner (2)
P. Morris (1)
D. Nill (3)
G. Storch (1)
M.D.Tuttle (13)

BLACK AND WHITE PHOTOGRAPHS
J. Cerveny (1)
J. Gebhard (2)
E. Grimmberger (66)
H. Hackenthal (3)
O. v. Helversen (1)

BLACK AND WHITE DRAWINGS AND SONOGRAMS
T. Schneehagen

INDEX

LITERATURE
Barbour, R. W. & W. H. Davis. 1969. *Bats of America.* University Press of Kentucky, Lexington.

Fenton, M. B. 1983. *Just Bats.* University of Toronto Press, Toronto.

Hill, J. E. & J. D. Smith. 1984. *Bats: A Natural History.* University of Texas Press, Austin.

Schober, W. 1984. *The Lives of Bats.* Arco Publishing, New York.

Tuttle, M. D. 1988. *America's Neighborhood Bats.* University of Texas Press, Austin.

Tuttle, M. D. 1995. Saving North America's Beleaguered Bats. *National Geographic,* 188(3):37-57.

Whittaker, J. O. Jr. 1992. *The Audobon Society Field Guide to North American Mammals.* A. A. Knopf, New York.

GERMAN LITERATURE
AHLEN, I.: Identification of Scandinavian bats by their sounds. Swed. Univ. Agric. Sci., Dept. of Wildlife Ecology, Report 6, Uppsala 1981

ANDERA, M. UND HORACEK, I.: Poznavame nase savce (Wir bestimmen unsere Saugetiere). Mlada Fronta, Praha 1982

BAUER, K. UND SPITZENBERGER, F.: Rote Liste seltener und gefahrdeter Saugetierarten Osterreichs (Mammalia). In: Rote Listen gefahrdeter Tiere Osterreichs. Bundesministerium fur Gesundheit und Umweltschutz, Wien 1983

BAUMANN, F.: Die freilebenden Saugetiere der Schweiz. Bern 1949

BLAB, J.: Grundlagen fur ein Fledermaus-Hilfsprogramm.Themen der Zeit, 5, KildaVerlag, Greven 1980

BLAB, J., NOWAK, E., TRAUTMANN, W. UND SUKOPP, H.: Rote Listen der gefahrdeten Tiere und Pflanzen in der Bundesrepublik Deutschland. Naturschutz aktuell Nr. 1; 4. Aufl., Kilda-Verlag, Greven 1984

BRINK, F. H. VAN DEN: Die Saugetiere Europas. Verl. P. Parey, Hamburg u. Berlin 1975

BROSSET, A.: La Biologie des Chiropteres. Masson et Cie . Paris 1966

CERVENY, J.: Abnormal Coloration in Bats (Chiroptera) of Czechoslovakia.

Nyctalus (N.F.) 1,193-202 (1980)

CORBET, G. UND OVENDEN D.: Pareys Buch der Saugetiere. Alle wildlebenden Saugetiere Europas. Verl. P. Parey Hamburg u. Berlin 1982

DEBLASE, A.F.: The bats of Iran: Systematics, Distribution, Ecology. Fieldiana Zoology, New Series No. 4, Chicago 1980

EISENTRAUT, M.: Die Fledertiere. In: Grzimeks Tierleben, Bd. 11. Kindler Verlag, Zurich 1969

GAISLER, J.: Ecology of bats. In: Stoddart, M. (Ed.) Ecology of small mammals. Chapman and Hall Ltd., London 1979

GEBHARD, J.: Unsere Fledermause. 2. Aufl. Naturhistor. Museum Basel 1985

GRIMMBERGER, E. UND BORK H.: Untersuchungen zur Biologie, Okologie und Populationsdynamik der Zwergfledermaus, *Pipistrellus p. pipistrellus* (Schreber 1774) in einer groBen Population im Norden der DDR. Nyctalus (N.F.) 1, 55-73 u. 122-136 (1978)

HACKETHAL, H.: Zur Merkmalsvariabilitat mitteleuropaischer Bartfledermause unter besonderer Berucksichtigung der Verbreitung und der okologischen Anspruche von *Myotis brandti* (Eversmann 1845). Nyctalus (N.F.) 1, 293410 (1982)

HACKETHAL, H.: Fledermause. In: Stresemann. Exkursionsfauna, Bd. 3 Wirbeltiere. volk und Wissen VE Verlag, Berlin 1983

HAENSEL, J. UND NAFE, M.: Anleitungen zum Bau von Fledermauskasten und bisherige Erfahrungen mit ihrem Einsatz. Nyctalus (N. F.) 1 327-348 (1982)

HANAK, V. AND HORACEK, I.: Some comments of the taxonomy of *Myotis daubentoni* (Kuhl, 1819) (Chiroptera Mammalia). Myotis 21-22,719 (1983-1984)

HEISE, G.: Zur Fortpflanzungsbiologie der Rauhhautfledermaus *(Pipistrellus nathusit).* Nyctalus (N. F.) 2,1 - 15 (1984)

HEISE, G. UND SCHMIDT, A.: Wo uberwintern im Norden der DDR beheimatete Abendsegler *(Nyctalus noctula)?* Nyctalus (N.F.) 1, 81-84 (1979)

HEU, R.: Mammiferi d'Europa. Arnoldo Mondadori Editore, Milano 1968

HILL, J.E. AND SMITH, J.D.: Bats. A natural history. British. Museum, London 1984

HORACEK, J. AND HANAK, V. Generic status of *Pipistrellus savii* and comments on classification of the genus *Pipistrellus (Chiroptera, Vespertilionidae)*. MYOTIS 23-24, 9-16 (1986).

KONIG, C.: Wildlebende Saugetiere Europas. Verl. Ch. Belser, Stuttgart 1976

KOLB, A.: Die Geburt einer Fledermaus. Image 49, 5-13 (1972)

KULZER, E.: Die Herztatigkeit bei lethargischen und winterschlafenden Fledermausen. Zeitschr. f. vergl. Physiologie 56, 63-94 (1967)

KULZER, E.: Winterschlaf. Stuttg. Beitr. z. Naturk., Ser. C, H. 14, Staatl. Museum f. Naturkunde, Stuttgart 1981

KURSKOV, A.N.: Bats of Belorussia. ,,Nauka i Technika", Minsk 1981 (russisch)

KUZYAKIN, A. P.: Letuchie myshi (Fledermause). Izd. Sovetskaya Nauka, Moskva 1950 (russisch)

MASING, M.: Lendlased (Fledermause). ,,Valgus" Tallin 1984 (estnisch)

MOHRES, P.: Bildhoren - eine neuentdeckte Sinnesleistung der Tiere. Umschau, 60, 673-678 (1960)

NATUSCHKE, G.: Heimische Fledermause. Neue Brehm Bucherei H. 269. A. Ziemsen Verl. Wittenberg 1960

NEUWEILER, G.: Die Ultraschalljager. GEO, Nr. 1, (Hamburg) 98-113 (1981)

NEUWEILER, G.: Echoortung S. 708-722. In: Hoppe, W. et al. (Herausg.): Biophysik, 2. Aufl., Springer Verl., Berlin Heidelberg, New York 1982

PELIKAN, J., GAISLER, J. und RODL, P.: Nasi savci. Academia, Praha 1979

PUCEK, Z. (Ed.): Keys to Vertebrates of Poland. Mammals. PWN - Polish Scientific Publishers, Warszawa 1981

ROER, H.: Zur Bestandsentwicklung einiger Fledermause in Mitteleuropa. Myotis 18-19, 60-67 (1980 -1981)

ROER, H. (Herausg.): Berichte und Ergebnisse von Markierungsversuchen an Fledermausen in Europa. Decheniana, 18. Beihefi, Bonn 1971

RYBERG, O.: Studies on bats and bat parasites, Svensk Nature, Stockholm 1947

SCHILLING, D., SINGER, D. UND DILLER, H.: Saugetiere. BLV-Verlagsgesellschaft Munchen, Wien, Zurich 1983

SCHMIDT, U.: Vampirfledermause. Neue Brehm Bucherei H. 515. A. Ziemsen Verl. Wittenberg 1978

SCHROPFER, R., FELDMANN R. UND VIERHAUS, H. (Herausg.): Die Saugetiere Westfalens. Westfalisches Mus. f. Naturk., Landschafisverband Westfalen-Lippe, Munster 1984

SCHOBER, W.: Mit Echolot und Ultraschall. Die phantastische Welt der Fledertiere. Herder Verl., Freiburg 1983

SOKOLOV, I. I. (Herausg.) Die Saugetierfauna der UdSSR (russ.) Teil 1, Verl. `d. Akademie der Wissenschaflen d. UdSSR, Moskau - Leningrad 1963

TOSCHl, A. i LANZA B.: Fauna d'Italia. Vol. IV: Mammalia. Edizioni Calderini, Bologna 1959

TUPINIER, Y.: Description d une chauve-souris nouvelle: *Myotis nathalinae* nov. spec. (Chiroptera-Vespertilionidae). Mammalia 41, 327-340 (1977)

YALDEN, B.W. AND MORRIS P. A.: The Lives of Bats. David and Charles Newton, Abbot, London, Vancouver 1975
--- Erste Durchfuhrungsbestimmung zur Naturschutzverordnung
--- Schutz von Pflanzen und Tierarten -(Artenschutzbestimmung) vom 1. Oktober 1984. Gesetzblatt der Deutschen Demokratischen Republik, Teil I, Nr. 31. Berlin 1984

Fledermauszeitschriften:
,,MYOTIS" - Herausgeber: Zoolog. Forschungsinstitut u. Museum A. Koenig BRD - 5300 Bonn.
,,NYCTALUS" - Herausgeber: Tierpark Berlin (Prof Dr. Dr. Dathe). DDR - 1136 Berlin
,,BAT NEWS" - Herausgeber: Fauna and Flora Preservation Soc., c/o Zoolog. Soc. of London. Regents Park, London NW 14 RY (UK).
,,BATS" - Herausgeber: Bat Conservation International P. O. Box 162 603 Austin, Texas 78716-2603 USA

MOST IMPORTANT MEASUREMENTS

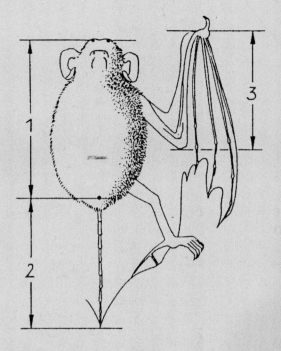

1: Head-body Length
2: Tail Length
3: Forearm Length

4: Ear Length (ear)
5a: Tragus Length
5b: Tragus Width

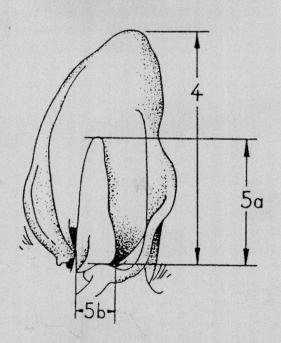

6a: Thumb Length
6b: Length of the thumb Claw

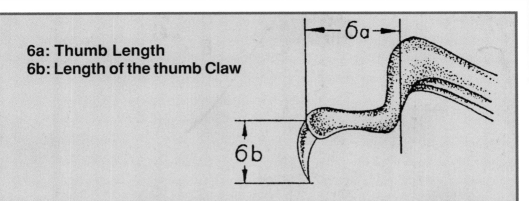

7: Wingspan
8: Length of the Fifth Finger (more correctly: length of the fifth metacarpal and the fifth finger)

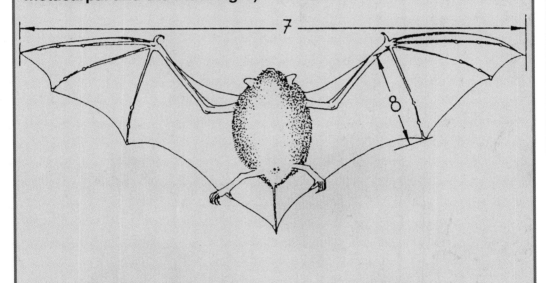

9: Condylobasal Length

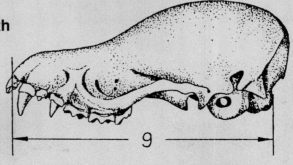

European and North
American Bats
TS-289